Springer Proceedings in Business and Economics

Springer Proceedings in Business and Economics brings the most current research presented at conferences and workshops to a global readership. The series features volumes (in electronic and print formats) of selected contributions from conferences in all areas of economics, business, management, and finance. In addition to an overall evaluation by the publisher of the topical interest, scientific quality, and timeliness of each volume, each contribution is refereed to standards comparable to those of leading journals, resulting in authoritative contributions to the respective fields. Springer's production and distribution infrastructure ensures rapid publication and wide circulation of the latest developments in the most compelling and promising areas of research today.

The editorial development of volumes may be managed using Springer's innovative Online Conference Service (OCS), a proven online manuscript management and review system. This system is designed to ensure an efficient timeline for your publication, making Springer Proceedings in Business and Economics the premier series to publish your workshop or conference volume.

More information about this series at http://www.springer.com/series/11960

Arnab Kumar Laha
Editor

Applied Advanced Analytics

6th IIMA International Conference on Advanced Data Analysis, Business Analytics and Intelligence

Springer

Editor
Arnab Kumar Laha
Production and Quantitative Methods
Indian Institute of Management Ahmedabad
Ahmedabad, Gujarat, India

ISSN 2198-7246 ISSN 2198-7254 (electronic)
Springer Proceedings in Business and Economics
ISBN 978-981-33-6655-8 ISBN 978-981-33-6656-5 (eBook)
https://doi.org/10.1007/978-981-33-6656-5

This Springer imprint is published by the registered company Springer Nature Singapore Pte Ltd.
The registered company address is: 152 Beach Road, #21-01/04 Gateway East, Singapore 189721, Singapore

Contents

Editor and Contributors

About the Editor

Arnab Kumar Laha is Professor of Production and Quantitative Methods (P&QM) at Indian Institute of Management Ahmedabad. He takes a keen interest in understanding how analytics, machine learning, and artificial intelligence can be leveraged to solve complex problems of business and society. He has published his research in national and international journals of repute, authored a popular book on analytics and an edited book volume published by Springer. He has been named as one of the "20 Most Prominent Analytics and Data Science Academicians in India" by Analytics India Magazine in 2018.

Contributors

Rishabh Agrawal Supply Chain and Operations Analytics, Applied Intelligence, Accenture Digital, Gurugram, India

Kingshuk Banerjee Cognitive Computing and Analytics, IBM Global Business Services, Bengaluru, India

Geetika Bansal Experian Credit Information Company of India, Mumbai, India

Lakshay Bansal Vinod Gupta School of Management, Indian Institute of Technology Kharagpur, Kharagpur, India

Rahul Bhattacharya Department of Statistics, University of Calcutta, Kolkata, India

Amol A. Bhave Cognitive Business and Decision Support, IBM India, Bengaluru, India

Atanu Biswas Applied Statistics Unit, Indian Statistical Institute, Kolkata, India

Saurav Chakravorty Brillio Technologies, Bengaluru, India

Abhijeet Chandra Vinod Gupta School of Management, Indian Institute of Technology Kharagpur, Kharagpur, India

Seinjuti Chatterjee Unbxd, Bengaluru, India

Ramya Cherukupalli Deloitte Consulting, Hyderabad, India

Aditya Chourasiya Vinod Gupta School of Management, Indian Institute of Technology Kharagpur, Kharagpur, India

Kalyani Dacha Deloitte Consulting, Hyderabad, India

Paulami Das Brillio Technologies, Bengaluru, India

Avijit Ghosh Vinod Gupta School of Management, Indian Institute of Technology Kharagpur, Kharagpur, India

Shyamal Ghosh Department of Mathematical Statistics and Actuarial Science, University of the Free State, Bloemfontein, South Africa

Abhishek Halder Tata Consultancy Services (TCS), Kolkata, India

Akshay Jain SVKM's Narsee Monjee Institute of Management Studies (NMIMS), Indore, India

Nishtha Jain Data Modeler, Experian Credit Information Company of India, Mumbai, India

Shrey Jain SVKM's Narsee Monjee Institute of Management Studies (NMIMS), Indore, India

Prasad Joshi Unbxd, Bengaluru, India

Ruhul Ali Khan Department of Mathematics, Indian Institute of Engineering Science and Technology, Howrah, India

Abhishek Kushwaha Brillio Technologies, Bengaluru, India

Arnab Kumar Laha Indian Institute of Management Ahmedabad, Ahmedabad, India

Priyanka Majumder Department of Mathematics, Indian Institute of Technology Bombay, Mumbai, India

Nitin Merh Jaipuria Institute of Management Indore, Indore, India

Ravi Shankar Mishra Unbxd, Bengaluru, India

Murari Mitra Department of Mathematics, Indian Institute of Engineering Science and Technology, Howrah, India

Biswaroop Mookherjee Tata Consultancy Services (TCS), Kolkata, India

Taranga Mukherjee Department of Statistics, University of Calcutta, Kolkata, India

Sandip Kumar Pal Cognitive Business and Decision Support, IBM India, Bengaluru, India

Shreya Piplani Experian Credit Information Company of India, Mumbai, India

Rudra P. Pradhan Indian Institute of Technology Kharagpur, Kharagpur, India

Sagar Raichandani Unbxd, Bengaluru, India

Aditi Singh Data Modeler, Experian Credit Information Company of India, Mumbai, India

Abir Sinha Deloitte Consulting, Hyderabad, India

Vibhu Srivastava Analytics Centre of Excellence, Max Life Insurance, New Delhi, India

R. Sujatha PSG Institute of Management, Coimbatore, India

R. P. Suresh Supply Chain and Operations Analytics, Applied Intelligence, Accenture Digital, Gurugram, India

Sanjay Thawakar Analytics Centre of Excellence, Max Life Insurance, New Delhi, India

B. Uma Maheswari PSG Institute of Management, Coimbatore, India

Shikha Verma Indian Institute of Management Ahmedabad, Ahmedabad, India

Machine Learning for Streaming Data: Overview, Applications and Challenges

Shikha Verma

This chapter gives a brief overview of machine learning for streaming data by establishing the need for special algorithms suited for prediction tasks for data streams, why conventional batch learning methods are not adequate, followed by applications in various business domains.

Section 1 gives a brief overview of machine learning definitions and terminologies with an emphasis on the challenges faced while mining streaming data. Section 2 gives a panoramic view of the literature on classification and regression for data streams. Section 3 gives a review of existing research on drift detection algorithms, both in supervised and unsupervised manner. Section 4 provides interesting application areas of the algorithms discussed in Sects. 3 and 4.

1 Introduction to Machine Learning and Streaming Data

The past decade has witnessed a rapid decline in cost of capturing, storing and analysing data which has facilitated a surge in interest in 'machine learning' by academics and practitioners alike. Static programming relies on manual effort in building an exhaustive set of conditions expected to be encountered by the programme and their associated actions. In contrast, machine learning algorithms minimise human effort by letting the data speak for itself by making the models learn from the data. The goal of a machine learning model is to learn from historic data and make predictions on unseen data. The learning phase is called the training phase, and the predicting phase is called the testing phase. Generalisability is an important model attribute that signifies that the model performance on training and test data is

---

S. Verma (✉)
Indian Institute of Management Ahmedabad, Ahmedabad, India
e-mail: phd16shikhav@iima.ac.in

A. K. Laha (ed.), *Applied Advanced Analytics*, Springer Proceedings in Business and Economics, https://doi.org/10.1007/978-981-33-6656-5_1

comparable; i.e., the learnings from training data are useful for predictions on unseen data.

Common machine learning algorithms include support vector machines, neural networks, and decision trees. Machine learning has been adopted widely in fields like marketing and sales, logistics, particle research, and autonomous mobility to name a few, and the business value created by machine learning and artificial intelligence is projected to reach $3.9 T in 2022 (Columbus 2019).

Overfitting and bias–variance tradeoff, especially in low sample size scenarios, are concerns in conventional machine learning. However, in this day and age, data size is huge, rather time and memory requirements for processing these huge volumes of data are the bottleneck (Domingos and Hulten 2000). The prevalence of information and communication technologies like the Internet of Things, GPS, and wearable sensors and mobile Internet devices in our everyday lives has heralded the age of big data—characterised by 3 V's: volume, variety, and velocity. Consequently, the development of efficient algorithms to mine big data has attracted significant attention in the literature. Streaming data is a variant of big data with a salient velocity aspect. Agarwal (2007) defines data streams as 'sequence of data arriving from a source at very high velocity'. Data streams are potentially infinite in nature, exerting high demands of disc storage.

Developing algorithms for streaming data is particularly challenging as a single pass at each incoming observation is allowed and the model has to be ready to predict anytime (Bifet et al. 2009).

2 Classification and Regression in Streaming Data

Supervised Learning can be viewed as a problem of predicting the output (dependent variable) based on some input variables (independent variables) which are measured and known (Hastie and Friedman 2003). The prediction task is termed as regression if the dependent variable is quantitative and classification if the dependent variable is qualitative.

While classification and regression have received ample attention in conventional machine learning literature, in this chapter, we will limit the view of these tasks from a stream learning perspective.

For both these tasks, the key challenges faced while analysing data streams are:

(a) Open-ended nature of data—Data streams from sensors are high velocity and potentially infinite in nature. This makes the storage of their original representation on a disc virtually impossible and creates a need for concise representations of the stream obtained by forgetting techniques such as sliding windows, the exponential weighting of old observations, sampling and forming synopsis structures such as histograms and wavelets.

(b) Concept drift—Due to the dynamic operating environment, the nature of data evolves over time leading to the model built on earlier data obsolete.

(c) Stability-Plasticity dilemma—forming new models at very short intervals can help the model 'track' the changes in data effectively but introduce instability in the system and is computationally expensive, especially when nature and timing of drift are unknown. On the other extreme, not updating the model at all can make it obsolete for predictions in the new data environment. Finding the right frequency of model update is a critical decision while designing machine learning models for streaming data

2.1 *Classification Algorithms*

Domingos and Hulten (2000) propose an incremental learning algorithm for decision trees that can learn optimal splits on multiple attributes based on information theoretic criterion (Gini index, etc.) from small samples only as entire volumes of new portions of streams are infeasible to process. They use Hoeffding bounds to ascertain that the splitting choice of trees formed on incremental data and batch data are asymptotically similar.

Ensemble machine learning models have often shown superior performance to monolithic models in predictive tasks as the overall learning process is more diverse and adaptable to multiple drift properties evolving over time (Khamassi et al. 2018). Extending this result to streaming data, Street and Kim (2001) introduced an ensemble learning algorithm (SEA) using decision trees. It involves continuous retraining of classifiers on incoming, new data chunks, and training of new classifiers which are only added to the predictive ensemble if they improve the overall prediction, and in lieu, a poorly performing classifier is dropped from the ensemble. Developing this idea further, Kotler and Maloof (2003) created a weighted ensemble where classifiers are given weights dynamically or dropped from ensemble based on their deviation from the global ensemble prediction.

Kourtellis et al. (2016) propose a distributed algorithm for learning decision trees from massive amounts of data arriving at high velocity which enables parallel computation and improves scalability on real-life datasets. Interested readers are referred to Parthasarathy et al. (2007) for a detailed review of distributed mining methods for data streams.

Many times, in classification problems encountered in real life, we witness 'class evolution', i.e. emergence of a new class and change in prior distributions of older classes. Some examples can be emergence of new topics on social media chatter in the natural language processing task of topic classification and introduction of a new transport mode in the task of transport mode detection. For detecting concept drift in such cases, Sun et al. (2016) propose an ensemble approach where the base learner for each class is updated with new data to adapt to changing class distributions.

2.2 *Regression Algorithms*

The Hoeffding bounds on classifier performance of very fast decision trees (VFDT) proposed by Domingos and Hulten (2000) can also be extended to regression problems. Ikonomovska and Gama (2008) propose fast and incremental regression trees (FIRT) using the Chernoff bound to determine the necessary sample size for a particular splitting variable. After obtaining a split, the authors propose training neural networks in incremental fashion in leaves of the node. The method is demonstrated to have low time and space complexity, works well with numerical attributes, and is robust to noise. FIRT with explicit drift detection capability based on sequential statistical tests was introduced by Ikonomovska et al. (2015) which only allows for model adaption when change is detected, saving on global model adaptation costs when incoming data is sufficiently stable. Ikonomovska et al. (2015) propose methods to form ensembles of Hoeffding trees for regression in evolving data streams.

3 Drift Detection Algorithms

As discussed in the previous sections, due to the changing nature of data, there is a need for models trained on initial data chunk to attune themselves to the most recent data concepts. Broadly categorising, there are two ways to do that: active and passive. Active drift detection methods explicitly detect drift based on incoming data characteristics, discard old data, and retrain classifier if and only if a drift is detected which is a computationally parsimonious way to adapt to concept drift.

Passive methods update model irrespective of drift by retraining on new data as it arrives or in chunks/windows. Sliding window models, exponential weighting of older observations, and dropping the weakest learner in constantly evolving ensemble are passive methods of adapting the model to recent changes in data distribution. As is evident, they are computationally very expensive, require constant access to new labels, and only suitable for small-scale problems. In the following subsections, we will limit the discussion to active drift detection methods only.

Concept drift can occur in multiple ways as real-life data streams are generated via complex, dynamic processes. Virtual drift refers to a scenario where predictor distribution; i.e. $p(x)$ changes but $p(y|x)$ remains unchanged. Real drift refers to a scenario where conditional posterior $p(y|x)$ changes.

In addition to its nature, concept drift can also be characterised in several other ways: speed (abrupt/gradual), severity (global/local), and recurrency (cyclic/acyclic). In real-life scenarios, drift detection problem is particularly challenging as multiple drift characteristics are evolving simultaneously (Khamassi et al. 2018). Sections 3.1 and 3.2 give an overview of supervised and unsupervised drift detection algorithms.

3.1 Supervised Drift Detection

Supervised drift detection refers to the task of drift detection when labels of all successive observations in the stream are known and instantaneously available. In one of the initial research efforts in this area, Page (1954) proposes a cumulative sum (CUSUM) which use deviation from original mean as a measure for drift detection and perform well for univariate sequence data.

Gama et al. (2004) propose a drift detection method (DDM) that monitors the error rate of learning algorithm over new training data. If the error rate increases on a temporal scale, it indicates a likely change in underlying distribution of data. As error computation requires continuous availability of predicted and actual labels, this method is supervised in nature. While DDM exhibits good performance in detecting sudden drifts, its performance in detecting gradual drift was superseded by early drift detection method (EDDM) proposed by Baena-Garcia et al. (2006).

Windowing has been used extensively in streaming data literature to update the model as per the latest data concept. However, deciding on the window size a priori is as challenging as it is important as window size represents a tradeoff between sensitivity and stability. Bifet and Gavalda (2007) propose adaptive windowing (ADWIN) algorithm which computes summary statistics on data windows and uses them as inputs to contract the window size when a 'sufficiently big' change is detected and expand the window size when no change is detected.

Bifet et al. (2009) introduce the adaptive size Hoeffding tree (ASHT), a method that entails building an ensemble of trees of varying sizes to adapt to gradual and abrupt drifts, with each tree having a weight inversely proportional to its error rate. The authors extended the idea of incremental learning in Hoeffding trees with mathematical guarantees on performance to an ensemble setup which has more diverse learners and better scalability.

As true labels are available, the nature of supervised drift detection problem has a high resemblance to a typical classification task, Therefore, the evaluation metrics used for supervised drift detection are accuracy, recall, precision, specificity, and sensitivity.

3.2 Unsupervised Drift Detection

Unsupervised drift detection algorithm broadly uses concepts of clustering to summarise trends in data and detect 'fading' and 'evolving' trends on a temporal scale. Spinosa et al. (2007) propose a novelty detection algorithm which is used to indicate drift in data sequences. The authors use K-means clustering algorithm on existing data to define a boundary for the 'normal' concept. If a sufficient number of new, incoming observations fall outside the normal concept boundary, they are labelled as a novel concept. Various adaptations of clustering methods are used by

Masud et al. (2010), Faria et al. (2013), Sethi et al. (2016) for novelty and concept drift detection.

Clustering methods have been used to detect novelties/drifts not only in sequence data but also in graph data. Aggarwal and Yu (2005) use clustering techniques on graph data to detect expanding and contracting communities in graphs based on changing interaction patterns. The use cases of this algorithm can be in identifying evolving user subgroups based on phone calls references between individuals, changing lending patterns in P2P microfinance groups, and information diffusion in social media communities.

The evaluation metrics for an unsupervised drift detection task are false alarms, sensitivity to drift, and additionally computation time for real-time applications.

4 Applications

This section gives a brief overview of the motivation and the present state of research of machine learning for data streams in various application areas such as recommender systems, transport, and mobility and fraud detection.

(a) Recommender systems
Recommender systems have been used extensively by e-commerce firms to reduce the information overload on customers due to the presence of millions of products on their platforms by analysing the user's past behaviour on the platform and suggesting products they are most likely to find valuable (Bobadilla et al. 2013). However, user preferences change over time due to evolving market dynamics and introduction of new products. Traditional approaches to recommender systems do not incorporate this temporal dynamics and model item user relationship in a static way, akin to batch learning. Nasraoui et al. (2007) study the behaviour of collaborative filtering-based recommender systems in streaming setup with concept evolution and found its performance to be inferior to that in static scenarios. Chang et al. (2017) propose a streaming recommender system which models user ratings over item space as time dependent and uses recursive mean-field approximation to estimate the updated user preference.

(b) Mobility and Transportation
The pervasive use of GPS sensors across transport modes has enabled accurate records of vehicle trajectories. This is leveraged by transport authorities and researchers for short- and long-term demand estimation, real-time traffic management, vehicle routing, and mode split estimation (Bajwa et al. 2011; Huang et al. 2019). However, mobility patterns change dynamically due to multiple factors such as pricing, service quality and quantity, demographic and socio-economic factors, and land use patterns. Models built in streaming setup are more suited for prediction and classification tasks in such a context. Laha and Putatunda (2018) propose regression models and their ensembles in sliding window setup with exponential fading strategy to predict drop-off location based

on location data stream of ongoing trip in GPS taxis. The models are demonstrated to have superior predictive performance with respect to their counterparts in batch setup considering accuracy and computation time tradeoff.
Moreira et al. (2013) propose streaming data models to predict taxi demand over spatial grids in the city over short horizons. Accurate forecasts of short-term demand helps in efficient vehicle routing and better fleet management in order to meet customer demand faster.
Boukhechba et al. (2015) propose association rule mining on GPS trajectories of individuals to learn their mobility habits and use this knowledge to predict the next location likely to be visited by them.

(c) Anomaly detection

Anomaly detection involves uncovering unusual and exceptional patterns from recognised patterns. Email spam filtering, fraud detection in banking, and finance and intrusion detection in networks are examples of anomaly detection tasks (Chandola et al. 2009; Mazhelis and Puuronen 2007).

Anomaly detection models are trained on historical data but spam and intrusion patterns evolve over time, created a need for adaptive models. Hayat et al. (2010) propose an adaptive language model for spam filtering with concept drift detection abilities using KL divergence. Parveen et al. (2011) use ensembles of adaptive, multiple classification models to identify malicious inside users trying to gain unauthorised access to information from historical activity log data.

5 Conclusion

In this chapter, we discussed the challenges faced while extracting intelligence from data streams and give an overview of key predictive algorithms and drift detection methods present in literature.

However, there is a need to develop adaptive algorithms as per application type and keeping real-life challenges in time such as limited and costly access to ground truth labels and computation time and accuracy tradeoff, especially for real-time decision-making tasks. Also, as hardware becomes increasingly inexpensive and deployment of machine learning models becomes pervasive in multiple domains, the main concerns are about using the computing resources judiciously. This has opened up an exciting new area of research called 'green machine learning' which aims to improve algorithmic efficiency by reducing the computing resources required and consequently minimise the carbon footprint of data centres and personal machines running machine learning models (Strubell et al. 2019). To sum it up, as the digital world generates ever increasing amounts of data, machine learning for streaming data comes with its own set of challenges but has immense potential of delivering business value.

References

Aggarwal, C. C. (Ed.). (2007). *Data streams: Models and algorithms* (Vol. 31). Springer Science & Business Media.

Aggarwal, C. C., & Yu, P. S. (2005, April). Online analysis of community evolution in data streams. In *Proceedings of the 2005 SIAM International Conference on Data Mining* (pp. 56–67). Society for Industrial and Applied Mathematics.

Baena-Garcia, M., del Campo-Ávila, J., Fidalgo, R., Bifet, A., Gavalda, R., & Morales-Bueno, R. (2006, September). Early drift detection method. In *Fourth international workshop on knowledge discovery from data streams* (Vol. 6, pp. 77–86).

Bajwa, R., Rajagopal, R., Varaiya, P., & Kavaler, R. (2011, April). In-pavement wireless sensor network for vehicle classification. In *Proceedings of the 10th ACM/IEEE International Conference on Information Processing in Sensor Networks* (pp. 85–96). IEEE.

Bifet, A., & Gavalda, R. (2007, April). Learning from time-changing data with adaptive windowing. In *Proceedings of the 2007 SIAM international conference on data mining* (pp. 443–448). Society for Industrial and Applied Mathematics.

Bifet, A., Holmes, G., Pfahringer, B., Kirkby, R., & Gavaldà, R. (2009, June). New ensemble methods for evolving data streams. In *Proceedings of the 15th ACM SIGKDD international conference on Knowledge discovery and data mining* (pp. 139–148). ACM.

Bobadilla, J., Ortega, F., Hernando, A., & Gutiérrez, A. (2013). Recommender systems survey. *Knowledge-based systems, 46*, 109–132.

Boukhechba, M., Bouzouane, A., Bouchard, B., Gouin-Vallerand, C., & Giroux, S. (2015). Online prediction of people's next Point-of-Interest: Concept drift support. In *Human Behavior Understanding* (pp. 97–116). Springer, Cham.

Chandola, V., Banerjee, A., & Kumar, V. (2009). Anomaly detection: A survey. *ACM Computing Surveys (CSUR), 41*(3), 15.

Chang, S., Zhang, Y., Tang, J., Yin, D., Chang, Y., Hasegawa-Johnson, M. A., & Huang, T. S. (2017, April). Streaming recommender systems. In *Proceedings of the 26th International Conference on World Wide Web* (pp. 381–389). International World Wide Web Conferences Steering Committee.

Columbus, L. (2019, March). Roundup of Machine Learning Forecasts and Market Estimates for 2019. *Forbes*. Retrieved from https://www.forbes.com/sites/louiscolumbus/2019/03/27/roundup-of-machine-learning-forecasts-and-market-estimates-2019/#206e54247695

Domingos, P. M. (2012). A few useful things to know about machine learning. *Communications of the ACM, 55*(10), 78–87.

Domingos, P., & Hulten, G. (2000, August). Mining high-speed data streams. In *Kdd* (Vol. 2, p. 4).

Faria, E. R., Gama, J., & Carvalho, A. C. (2013, March). Novelty detection algorithm for data streams multi-class problems. In *Proceedings of the 28th annual ACM symposium on applied computing* (pp. 795–800). ACM.

Gama, J., Medas, P., Castillo, G., & Rodrigues, P. (2004, September). Learning with drift detection. In *Brazilian symposium on artificial intelligence* (pp. 286–295). Springer, Berlin, Heidelberg.

Hastie T. T. R., & Friedman, J. H. (2003). Elements of statistical learning: data mining, inference, and prediction.

Hayat, M. Z., Basiri, J., Seyedhossein, L., & Shakery, A. (2010, December). Content-based concept drift detection for email spam filtering. In *2010 5th International Symposium on Telecommunications* (pp. 531–536). IEEE.

Huang, H., Cheng, Y., & Weibel, R. (2019). Transport mode detection based on mobile phone network data: A systematic review. *Transportation Research Part C: Emerging Technologies.*

Ikonomovska, E., & Gama, J. (2008, October). Learning model trees from data streams. In *International Conference on Discovery Science* (pp. 52–63). Springer, Berlin, Heidelberg.

Ikonomovska, E., Gama, J., & Džeroski, S. (2015). Online tree-based ensembles and option trees for regression on evolving data streams. *Neurocomputing, 150,* 458–470.

Ikonomovska, E., Gama, J., Sebastião, R., & Gjorgjevik, D. (2009, October). Regression trees from data streams with drift detection. In *International Conference on Discovery Science* (pp. 121–135). Springer, Berlin, Heidelberg.

Khamassi, I., Sayed-Mouchaweh, M., Hammami, M., & Ghédira, K. (2018). Discussion and review on evolving data streams and concept drift adapting. *Evolving Systems, 9*(1), 1–23.

Kolter, J. Z., & Maloof, M. A. (2003, November). Dynamic weighted majority: A new ensemble method for tracking concept drift. In *Third IEEE international conference on data mining* (pp. 123–130). IEEE.

Kourtellis, N., Morales, G. D. F., Bifet, A., & Murdopo, A. (2016, December). Vht: Vertical hoeffding tree. In *2016 IEEE International Conference on Big Data (Big Data)* (pp. 915–922). IEEE.

Laha, A. K., & Putatunda, S. (2018). Real time location prediction with taxi-GPS data streams. *Transportation Research Part C: Emerging Technologies, 92*, 298–322.

Laney, D. (2001). 3D data management: Controlling data volume, velocity and variety. *META Group Research Note, 6*(70), 1.

Masud, M., Gao, J., Khan, L., Han, J., & Thuraisingham, B. M. (2010). Classification and novel class detection in concept-drifting data streams under time constraints. *IEEE Transactions on Knowledge and Data Engineering, 23*(6), 859–874.

Mazhelis, O., & Puuronen, S. (2007, April). Comparing classifier combining techniques for mobile-masquerader detection. In *The Second International Conference on Availability, Reliability and Security (ARES'07)* (pp. 465–472). IEEE.

Moreira-Matias, L., Gama, J., Ferreira, M., Mendes-Moreira, J., & Damas, L. (2013). Predicting taxi–passenger demand using streaming data. *IEEE Transactions on Intelligent Transportation Systems, 14*(3), 1393–1402.

Nasraoui, O., Cerwinske, J., Rojas, C., & Gonzalez, F. (2007, April). Performance of recommendation systems in dynamic streaming environments. In *Proceedings of the 2007 SIAM International Conference on Data Mining* (pp. 569–574). Society for Industrial and Applied Mathematics.

Page, E. S. (1954). Continuous inspection schemes. *Biometrika, 41*(1/2), 100–115.

Parthasarathy, S., Ghoting, A., & Otey, M. E. (2007). A survey of distributed mining of data streams. In *Data Streams* (pp. 289–307). Springer, Boston, MA.

Parveen, P., Evans, J., Thuraisingham, B., Hamlen, K. W., & Khan, L. (2011, October). Insider threat detection using stream mining and graph mining. In *2011 IEEE Third International Conference on Privacy, Security, Risk and Trust and 2011 IEEE Third International Conference on Social Computing* (pp. 1102–1110). IEEE.

Sethi, T. S., Kantardzic, M., & Hu, H. (2016). A grid density based framework for classifying streaming data in the presence of concept drift. *Journal of Intelligent Information Systems, 46*(1), 179–211.

Spinosa, E. J., de Leon F de Carvalho, A. P., & Gama, J. (2007, March). Olindda: A cluster-based approach for detecting novelty and concept drift in data streams. In *Proceedings of the 2007 ACM symposium on Applied computing* (pp. 448–452). ACM.

Street, W. N., & Kim, Y. (2001, August). A streaming ensemble algorithm (SEA) for large-scale classification. In *Proceedings of the seventh ACM SIGKDD international conference on Knowledge discovery and data mining* (pp. 377–382). ACM.

Strubell, E., Ganesh, A., & McCallum, A. (2019). Energy and Policy Considerations for Deep Learning in NLP. *arXiv preprint* arXiv:1906.02243.

Sun, Y., Tang, K., Minku, L. L., Wang, S., & Yao, X. (2016). Online ensemble learning of data streams with gradually evolved classes. *IEEE Transactions on Knowledge and Data Engineering, 28*(6), 1532–1545.

Binary Prediction

Arnab Kumar Laha

1 Introduction

Binary prediction is one of the most widely used analytical techniques having far-reaching applications in multiple domains. In the business context, it is used to predict which loans are likely to default, which policyholders are likely to discontinue an insurance policy, which customers are likely to change their service provider, which customers are likely to buy a newly released book, which transactions are likely to be fraud, etc. Apart from business applications, the binary prediction problem arises routinely in medicine, e.g., to determine whether a person has a certain disease or not (Shilaskar and Ghatol 2013), chemistry (Banerjee and Preissner 2018) and many other fields. Because of the huge importance of the binary prediction problem, a number of methods have been developed over the years. The more well-known and widely used methods are linear discriminant analysis, logistic regression, random forest, support vector machines and k-nearest neighbors (see James et al. 2013 for an introduction to these methods).

In this article, we concentrate on the binary prediction task. We discuss the well-known logistic regression predictor and compare its performance with a relatively less widely used predictor—the maximum score predictor using two real-life datasets. The two datasets considered in this paper are both unbalanced with one class having significantly larger number of observations than the other class. The maximum score predictor discussed in this article is based on a modification of the maximum score estimator introduced in Manski (1975). It is observed that the maximum score predictor performs better than the logistic regression predictor for these two real-life datasets.

The article is structured as follows: In Sect. 2, we briefly discuss the logistic regression from a prediction perspective; in Sect. 3, we discuss the use of the logistic

A. K. Laha (✉)
Indian Institute of Management Ahmedabad, Ahmedabad, India
e-mail: arnab@iima.ac.in

A. K. Laha (ed.), *Applied Advanced Analytics*, Springer Proceedings in Business and Economics, https://doi.org/10.1007/978-981-33-6656-5_2

regression for binary prediction; in Sect. 4, we introduce the maximum-score predictor; in Sect. 5, we compare the performance of the logistic regression predictor and the maximum-score predictor using two real-life datasets, and in Sect. 6 we make some concluding remarks.

2 Logistic Regression

In binary classification problems, the response variable (Y) is dichotomous (i.e., takes only two values which are coded as 0 and 1). The predictor variables are typically either numeric or categorical though other types of variables have also been considered in the academic literature. We assume that we have k predictor variables $X_1, \ldots, X_k$ which are all numeric. In the logistic regression model, we try to fit a model

$$P(Y = 1) = \frac{e^{\beta_0+\beta_1 X_1+\ldots+\beta_k X_k}}{1 + e^{\beta_0+\beta_1 X_1+\ldots+\beta_k X_k}}$$

where $\beta_0, \ldots, \beta_k$ are unknown constants that have to be estimated from the given data.

Let $(y_i, x_{1i}, \ldots, x_{ki}), i = 1, \ldots, n$ be a random sample of size n from the target population. Then,

$$P(Y_i = y_i) = \left(\frac{e^{\beta_0+\beta_1 x_{1i}+\ldots+\beta_k x_{ki}}}{1 + e^{\beta_0+\beta_1 x_{1i}+\ldots+\beta_k x_{ki}}}\right)^{y_i} \left(1 - \frac{e^{\beta_0+\beta_1 x_{1i}+\ldots+\beta_k x_{ki}}}{1 + e^{\beta_0+\beta_1 x_{1i}+\ldots+\beta_k x_{ki}}}\right)^{1-y_i}$$

where $y_i = 0$ or 1.

The parameters of the logistic regression model are estimated using the maximum likelihood estimation (MLE) method. The likelihood is

$$L(\beta_0, \ldots, \beta_k) = \prod_{i=1}^{n} \left(\frac{e^{\beta_0+\beta_1 x_{1i}+\ldots+\beta_k x_{ki}}}{1 + e^{\beta_0+\beta_1 x_{1i}+\ldots+\beta_k x_{ki}}}\right)^{y_i} \left(1 - \frac{e^{\beta_0+\beta_1 x_{1i}+\ldots+\beta_k x_{ki}}}{1 + e^{\beta_0+\beta_1 x_{1i}+\ldots+\beta_k x_{ki}}}\right)^{1-y_i}$$

The values of $\beta_0, \ldots, \beta_k$ for which $L(\beta_0, \ldots, \beta_k)$ is maximized are the MLEs, and these are denoted as $\hat{\beta}_0, \ldots, \hat{\beta}_k$. Given a new observation for which the values of the predictor variables are known, say $(x_1^*, \ldots, x_k^*)$, but the value of the response variable Y^* is unknown we can estimate

$$P(Y^* = 1) = \hat{p} = \frac{e^{\hat{\beta}_0+\hat{\beta}_1 x_1^*+\ldots+\hat{\beta}_k x_k^*}}{1 + e^{\hat{\beta}_0+\hat{\beta}_1 x_1^*+\ldots+\hat{\beta}_k x_k^*}}$$

The delta method (Small 2010) can be used to obtain the approximate standard error of the estimated probability when the sample size n is large. Let $\boldsymbol{\beta}' = (\beta_0, \beta_1, \ldots, \beta_k)$ and $\mathbf{x}^{*\prime} = (1, x_1^*, \ldots, x_k^*)$, then in matrix notation we have $\hat{p} = \frac{e^{\mathbf{x}^{*\prime}\boldsymbol{\beta}}}{1+e^{\mathbf{x}^{*\prime}\boldsymbol{\beta}}}$. An application of delta method yields the estimated asymptotic standard error of $\hat{p}$ as $se(\hat{p}) = \hat{p}(1-\hat{p})\mathbf{x}^{*\prime}\hat{\mathbf{V}}(\hat{\beta})\mathbf{x}^*$ where $\hat{\mathbf{V}}(\hat{\beta})$ is the estimated variance–covariance matrix of the estimated coefficients $(\hat{\beta}_0, \hat{\beta}_1, \ldots, \hat{\beta}_k)$. An approximate 95% confidence interval for $\hat{p}$ can then be obtained as $(\hat{p} - 2\,se(\hat{p}), \hat{p} + 2\,se(\hat{p}))$.

3 Application to Binary Prediction

The logistic regression model provides an estimate of the probability $P(Y^* = 1)$. When a prediction of Y^* is desired, this information is converted to an estimate of Y^* by use of a threshold c on the magnitude of $\hat{p}$, i.e., $\hat{Y}^* = 1$ if $\hat{p} > c$ and is $= 0$ otherwise. In other words,

$$\hat{Y}^* = 1_{\hat{p}>c}$$

where 1_A is the indicator of the event A.

The value of $c = 0.5$ is sometimes used as in this case the estimated $P(Y^* = 1)$ is greater than $P(Y^* = 0)$. However, in many real-life business applications it is observed that the estimated value of $P(Y^* = 1)$ is much smaller than 0.5 for most (or sometimes all) values of the predictors within their observed ranges derived based on the training data. Using the threshold $c = 0.5$ would lead to prediction $\hat{Y}^* = 0$ in most/all cases. This often defeats the purpose of building the prediction model. Hence, a good choice of c is important for real-life applications.

The threshold value $\hat{c}$ is obtained by studying the variation of the performance of the logistic regression predictor for different values of c and choosing that value for which the predictor meets the business objective to the greatest extent.

Some of the popular measures that are useful for understanding the performance of a binary predictor are accuracy, sensitivity and specificity. Let the logistic regression predictor be used on a dataset of m observations for which the information about the predictors as well as the response is known. For each observation, the logistic regression model is used to estimate the probability that the response value is 1. Then, a threshold c is used to convert the estimated probabilities into predicted responses (i.e., 0 or 1) as discussed in the previous section. Now let m_{00} be the number of observations for which both the actual and predicted values of response are 0, m_{01} be the number of observations for which the actual response is 0 and predicted value of the response is 1, m_{10} be the number of observations for which the actual response is 1 and predicted value of the response is 0, and m_{11} be the number of observations for which both the actual and predicted values of response are 1. Note that $m_{00} + m_{01} + m_{10} + m_{11} = m$. The measure accuracy is defined as

$$\text{Accuracy} = \frac{m_{00} + m_{11}}{m}$$

and is often expressed as a percentage. While being a good measure in situations where the response is a balanced mix of 0s and 1s, accuracy can be a misleading measure if one of the classes is dominant. As an example, suppose it happens that there are only 5% observations in a dataset whose response is 1. In this case, a predictor that predicts all observations to be 0s would be 95% accurate but would not be able to correctly predict a single observation whose response is 1. To avoid such problems, it is important to look at performance measures that take a more granular view. The sensitivity of a binary predictor is its accuracy in predicting response 1, i.e.,

$$\text{Sensitivity} = \frac{m_{11}}{m_{10} + m_{11}}$$

and the specificity is the accuracy of the predictor in predicting response 0, i.e.,

$$\text{Specificity} = \frac{m_{00}}{m_{00} + m_{01}}$$

These are often expressed as percentages. An effective binary predictor should have both high sensitivity and specificity desirably close to 100%. However, it is generally not possible to have both specificity and sensitivity close to 100% when dealing with real-life datasets and therefore based on the application context a trade-off between sensitivity and specificity is carried out while choosing the threshold value c. Note that all the performance measures discussed until now are all dependent on the choice of the threshold value c. As the threshold value c is varied in the range $0 \leq c \leq 1$, we obtain a set of points (Specificity (c), Sensitivity (c)). The receiver operating characteristic (ROC) curve is a plot of $\left(1 - \text{Specificity}(c), \text{Sensitivity}(c)\right)$, $0 \leq c \leq 1$. The area under the ROC curve (AUC) is often used as a summary measure of binary predictor performance with its ideal value being 1.

In practical applications, it is advisable to determine c using a "validation" dataset that is separate from the training dataset to reduce the chance of overfitting. For this purpose at the initial stage itself, the given data is divided randomly into three parts, training, validation and test datasets containing $100\alpha\%$, $100\beta\%$ and $100(1 - \alpha - \beta)\%$ of the data where $0 < \alpha, \beta < 1$ and $0 < \alpha + \beta < 1$. A popular choice for (α, β) is $(0.7, 0.2)$. The test data is used to get an idea about the performance of the binary predictor with new data.

4 Maximum Score Predictor

It is easy to check that $\hat{p} > \hat{c} \iff \mathbf{x}'\hat{\beta} > ln(\frac{\hat{c}}{1-\hat{c}})$. Writing $\hat{\hat{\beta}}_0 = \hat{\beta}_0 - ln(\frac{\hat{c}}{1-\hat{c}})$ and $\hat{\hat{\beta}} = (\hat{\hat{\beta}}_0, \hat{\beta}_1, \ldots, \hat{\beta}_k)$, we can rewrite $\mathbf{x}'\boldsymbol{\beta} > ln(\frac{\hat{c}}{1-\hat{c}})$ as $\mathbf{x}'\hat{\hat{\beta}} > 0$. This suggests an alternative approach to the binary prediction problem, wherein we consider binary predictors of the form $\hat{Y}^* = 1_{\mathbf{x}'\boldsymbol{\beta}>0}$ and estimate the unknown parameters β by maximizing a "score function." The score function can be accuracy or can be a function of specificity and sensitivity as discussed below. Manski (1985) suggests maximizing the accuracy on the training data for estimating the parameter $\boldsymbol{\beta}$. Since

$\mathbf{x}'\boldsymbol{\beta} > 0 \Longleftrightarrow k\mathbf{x}'\boldsymbol{\beta} > 0$ for any constant $k > 0$, to ensure the identifiability of $\boldsymbol{\beta}$ it is restricted to have unit Euclidean norm, i.e., $||\boldsymbol{\beta}|| = 1$. Other "score functions" that may be considered are Youden's index which is *Sensitivity* − *(1* − *Specificity)* and the G-mean which is the geometric mean of the specificity and sensitivity. Note that a good binary predictor would have Youden's index and G-mean as high as possible. For both of these measures, the maximum possible value is 1.

5 Examples

In this section, we provide two examples based on real-life publicly available datasets. For Example 1, we analyze the Amazon books dataset from DASL (https://dasl.datadescription.com/datafile/amazon-books/). We aim to predict whether a book is paperback (P) or hardcover (H) based on the information about their list price, height, width and thickness given in the dataset. In Example 2, we analyze a telecom customer churn dataset (https://www.kaggle.com/mnassrib/telecom-churn-datasets) provided by Orange. Here, we aim to predict churn using the predictors: total day minutes, total evening minutes, total night minutes, total international minutes and number of customer service calls.

5.1 *Example 1*

The numbers of "P" and "H" in the given dataset are not balanced, and the ratio $P : H$ is roughly[1] 3:1. We split the given dataset into three parts: training (70%), validation(20%) and test(10%).[2] For the purpose of comparison, we use the same training dataset and test dataset for the logistic regression predictor and maximum score predictor. The validation dataset is not used when working with the maximum score predictor. For the logistic regression predictor, the validation data is used to determine the threshold in two different ways: (i) minimizing the misclassification error (where Misclassification error $= m_{01} + m_{10}$) and (ii) maximizing the G-mean. The sensitivity, specificity and the median G-mean value on the test data are noted for all the three methods, i.e., logistic regression with threshold chosen by minimizing the misclassification error (LR-Misclass), logistic regression with threshold chosen by maximizing the G-mean (LR-G-mean) and maximum-score method with G-mean score (Max-Score). This whole exercise is repeated 100 times, and then the median sensitivity, median specificity and the G-mean of the three methods on the test data are noted (see Table 1).

[1]Of the 318 observations in the dataset, 84 books were H and the rest were P.

[2]The number of observations in training, validation and test datasets was 223, 64 and 31, respectively.

Table 1 Comparison of performances of the LR-Misclass, LR-G-mean and Max-Score predictors on the Amazon books dataset

	Median sensitivity	Median specificity	Median G-mean
LR-Misclass	0.5	0.87	0.66
LR-G-mean	0.75	0.74	0.70
Max-Score	0.75	0.78	0.73

Table 2 Comparison of performances of the LR-Misclass, LR-G-mean and Max-Score predictors on the telecom churn dataset

	Median sensitivity	Median specificity	Median G-mean
LR-Misclass	0	1	0
LR-G-mean	0.49	0.48	0.49
Max-Score	0.51	0.50	0.50

It can be seen that the Max-Score predictor performs better than both the logistic regression-based predictors, LR-Misclass and LR-G-mean, in terms of the median G-mean.

5.2 *Example 2*

The telecom customer churn data consisted of 3333 observations of which 14.5% were churners and the rest were not churners. Of the several variables available in the dataset, we chose only the five variables mentioned earlier for this example. The same steps as those followed in Example 1 above were followed, and Table 2 gives the results.

It may be noted that the accuracy of the LR-Misclass predictor is the highest but is of no use as it always predicts every observation in the test dataset as belonging to the non-churner class. This is a typical problem when dealing with datasets in which one class has many more observations compared to the other class. In these situations, the trivial predictor which assigns every new observation to the majority class has high accuracy but no business relevance. The use of G-mean alleviates this problem to some extent. We find that the Max-Score predictor performs slightly better than the LR-G-mean predictor in terms of the median G-mean.

6 Conclusion

In this article, we discuss the predictive performance of the Max-Score predictor vis-a-vis the LR-misclass and LR-G-mean predictors when dealing with unbalanced datasets by analyzing two real-life datasets. The results suggest that the Max-Score predictor with the G-mean as the score has better predictive performance than the logistic regression-based predictors. This indicates the need for further studies, with both real-life and simulated datasets, to examine the efficacy of the Max-Score predictor and also its limitations.

References

Banerjee, P., & Preissner, R. (2018). Bittersweetforest: A random forest based binary classifier to predict bitterness and sweetness of chemical compounds. *Frontiers in Chemistry*, *6*, 93.

James, G., Witten, D., Hastie, T., & Tibshirani, R. (2013). An introduction to statistical learning: With applications in r. Springer.

Manski, C. (1975). Maximum score estimation of the stochastic utility model of choice. *Journal of Econometrics*, *3*(3), 205–228.

Manski, C. F. (1985). Semiparametric analysis of discrete response: Asymptotic properties of the maximum score estimator. *Journal of Econometrics*, *27*(3), 313–333.

Shilaskar, S., & Ghatol, A. (2013). Feature selection for medical diagnosis : Evaluation for cardiovascular diseases. *Expert Systems with Applications*, *40*(10), 4146–4153.

Small, C. G. (2010). *Expansions and asymptotics for statistics*. Chapman: Hall/CRC.

Reliability Shock Models: A Brief Excursion

Murari Mitra and Ruhul Ali Khan

1 Introduction

Failures of devices and systems, equipment or components thereof, materials and structures, machines and biological entities are broadly classified into two types of failure categories: (1) units fail by physical deterioration due to constant wear and tear, i.e., degradation failure; (2) failure due to shocks arriving randomly in time. Shocks may be fatal or one might have a cumulative damage model where a unit fails when the total damage due to shocks exceeds a certain critical threshold. However, it may well happen that failure is caused by a combination of both shocks and degradation. Such damage models are commonly applied to actual units that are working in industry, service, information and computers. Some typical examples are given in the following list.

1. A vehicle axle fails when the depth of a crack exceeds a certain threshold. In real-life situations, a train axle is often replaced after a specified number of revolutions or after a certain distance travelled (see Akama and Ishizuka 1995). A tyre on an automobile provides a similar example (see Bogdanoff and Kozin 1985; Gertsbakh and Kordonskiy 2012).
2. A battery or cell stores energy by chemical change and converts it to electricity. Energy of the battery is depleted by use and it becomes useless at the end of the chemical change (see Satow et al. 2000). This corresponds to the damage model by replacing shock with use and damage with oxidation or deoxidation.

M. Mitra (✉) · R. A. Khan
Department of Mathematics, Indian Institute of Engineering Science and Technology, Howrah, India
e-mail: murarimitra@yahoo.com

R. A. Khan
e-mail: ruhulali.khan@gmail.com

A. K. Laha (ed.), *Applied Advanced Analytics*, Springer Proceedings in Business and Economics, https://doi.org/10.1007/978-981-33-6656-5_3

3. The strength of a fibrous carbon composite is essentially determined by the strength of its constituent fibres. When a composite specimen is placed under tensile stress, the fibres themselves may break within the material. Such materials are broken based on cumulative damage (see Durham and Padgett 1997; Padgett 1998).
4. Garbage collection in a database system is a simple method to reclaim the location of active data because updating procedures typically reduce storage areas and worsen processing efficiency. To use storage areas more effectively and to improve processing efficiency, garbage collections are done at suitable times. Such a garbage collection model is also an example of damage model where shock corresponds to update and damage corresponds to garbage (see Nakagawa 2007).
5. Data in a computer system are frequently updated by addition or deletion of items and are stored in secondary media. However, data files are sometimes corrupted by several errors due to noises, human errors and hardware faults. The most dependable method to ensure the safety of data consists of taking backup copies at appropriate times. This corresponds to the damage model by replacing shock with update and damage with dumped files (see Nakagawa 2007).

Furthermore, as Nakagawa (2007) points out, damage models have been applied to crack growth models (Bogdanoff and Kozin 1985; Sobczyk and Trebicki 1989), to welded joints (Lukić and Cremona 2001), floating structures (Garbatov and Soares 2001), reinforced concrete structures (Petryna et al. 2002) and plastic automotive components (Campean et al. 2005). Such stochastic models of fatigue damage of materials were described in detail by Sobczyk and Spencer (2012). Failure mechanisms of damage models in engineering systems have been summarized in Dasgupta and Pecht (1991).

In this article, our goal is to present a brief introduction to the extensive area of shock model research in reliability theory. For limitations of space, we choose to confine ourselves to damage models in which the system under consideration is not subject to constant wear and tear and only shocks are responsible for its failure.

Suppose that a system is subjected to a sequence of shocks which arrive randomly in time according to a counting process $\{N(t) : t \geq 0\}$ and cause damage to the system. Suppose V_{k+1} is the interarrival time between the k-th and $(k+1)$-st shocks. Let $\bar{P}_k$ be the probability that the system survives the first k shocks, $k = 0, 1, 2, \ldots$, and suppose T denotes the time to failure. We assume that $\bar{P}_k$'s satisfy the following natural condition:

$$1 = \bar{P}_0 \geq \bar{P}_1 \geq \bar{P}_2 \geq \cdots$$

and that the expected number of shocks required to cause failure is finite, i.e. $\sum_{j=0}^{\infty} \bar{P}_j = B < \infty$. Note that the probability that failure is caused by the k-th shock is given by $p_k = \bar{P}_{k-1} - \bar{P}_k,\ \ k = 1, 2, \ldots$ and we define $p_0 = 1 - \bar{P}_0$. Then the probability that the device survives beyond time t is given by the survival function

$$\begin{aligned}\bar{H}(t) := P(T > t) &= P[\text{the device survives beyond time } t] \\ &= \sum_{k=0}^{\infty} P[\text{the device survives beyond time } t \mid N(t) = k]P[N(t) = k] \\ &= \sum_{k=0}^{\infty} P[N(t) = k]\bar{P}_k \end{aligned} \tag{1}$$

where $N(t)$ denotes the number of shocks to which the system has been exposed upto time t. This is the prototype reliability *shock model*. During the last few decades, many authors have studied this model extensively in several scenarios where shocks arrive according to various counting processes. Some of these are listed below:

1. *Homogeneous Poisson process* (HPP), i.e. a counting process null at the origin with independent and stationary increments where the probability of a shock occurrring in $(t, t + \Delta t]$ is $\lambda\Delta t + o(\Delta t)$, while the probability of more than one shock occurring in $(t, t + \Delta t]$ is $o(\Delta t)$. Here shocks arrive with a constant intensity λ and interarrival time V_k's are independent, identically distributed exponential random variables.
2. *Nonhomogeneous Poisson process* (NHPP), i.e., a counting process null at the origin with independent increments where the probability of a shock in occurring in $(t, t + \Delta t]$ is $\lambda(t)\Delta t + o(\Delta t)$, while the probability of more than one shock occurring in $(t, t + \Delta t]$ is $o(\Delta t)$. Here $\lambda(t)$ is the (integrable) intensity function of the process.
3. *Stationary pure birth process*, i.e., shocks occur according to a Markov process; given that k shocks have arrived in $(0, t]$, the probability of a shock occurring in $(t, t + \Delta t]$ is $\lambda_k \Delta t + o(\Delta t)$, while the probability of more than one shock occurring in $(t, t + \Delta t]$ is $o(\Delta t)$. Here the shocks are independent and are governed by a birth process with intensities $\lambda_k, k = 0, 1, 2, \ldots$. Note that V_{k+1} is exponentially distributed with mean $1/\lambda_k$ for $k = 0, 1, 2, \ldots$.
4. *Nonstationary pure birth process*, i.e. shocks occur according to a Markov process; given that k shocks have arrived in $(0, t]$, the probability of a shock occurring in $(t, t + \Delta t]$ is $\lambda_k\lambda(t)\Delta t + o(\Delta t)$, while the probability of more than one shock occurring in $(t, t + \Delta t]$ is $o(\Delta t)$.
5. *renewal process*; i.e. interarrival time between two consecutive shocks are independent and identically distributed random variables.

Before proceeding further, a motivating example would be appropriate. Consider an insurance company where claims arrive randomly in time. Claims can be interpreted as 'shocks' and the magnitude X_i of the i-th claim can be interpreted as the damage caused by the i-th shock. When the total claim Z exceeds a threshold Y (insurance company's capital), the company goes bankrupt. This cumulative damage model is illustrated in Fig. 1. This idea is applicable in various scenarios such as risk analysis, inventory control, biometry, etc., by simply considering appropriate analogues. The following table (Table 1) shows certain application areas with corresponding interpretations of the concepts of 'shock', 'device failure' and 'survival until time t'.

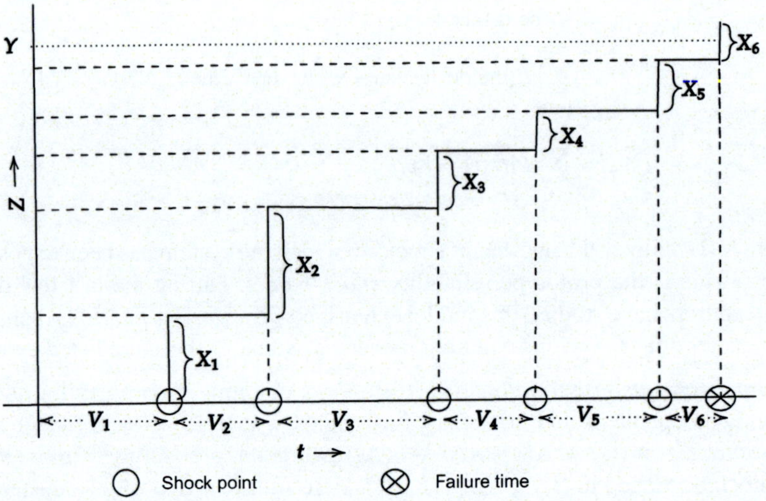

Fig. 1 Diagrammatic representation of a cumulative damage model

One may refer to Nakagawa (2007) and the references therein for a more extensive discussion of the application of shock model theory in diverse areas.

Using prior knowledge of the fact that the survival function $\bar{H}(t)$ belongs to a certain class of life distributions, it is feasible to obtain sharp bounds on the survival function and other parameters of the distribution. Such knowledge produces proper maintenance strategies in reliability contexts, higher accuracy in estimation results and functional financial policies in risk analysis. In the huge body of literature concerning shock model theory, the works of Shanthikumar and Sumita (1983), Anderson (1987), Yamada (1989), Kochar (1990), Pérez-Ocón and Gámiz-Pérez (1995), Pellerey (1993), Ebrahimi (1999) deserve special mention.

To the best of our knowledge, there is no review article available in the literature regarding nonparametric ageing classes arising from shock models. Reliability shock models are primarily concerned with the following fundamental question : if the sequence $(\bar{P}_k)_0^\infty$ possesses a certain discrete ageing property, does $\bar{H}(t)$ inherit the corresponding continuous version of the said ageing property under the trans-

Table 1 Analogues of shock model concepts

Reliability	Inventory control	Risk analysis	Biometry
Shock	Demand	Claim	Shock
Device failure	Shortage	Bankruptcy	Failure of organ or death of organism
Survival until time t	No shortage during $[0, t]$	All claims met during $[0, t]$	Survival until time t

formation (1) in various scenarios (i.e. when $N(t)$ is a HPP, NHPP, stationary or nonstationary pure birth process, etc.)? Before proceeding further, we introduce the concept of ageing and notions of ageing classes as these concepts are germane to the material that follows.

In the theory of reliability, survival or risk analysis, the concept of ageing plays a prominent role. The term '*ageing*' in the context of biological or mechanical devices defines how the residual life of the device is affected by its age in some probabilistic sense. '*No ageing*' means that the age of a component has no effect on the distribution of its residual lifetime. Positive ageing describes the situation where the residual lifetime tends to decrease, in some probabilistic sense, with increasing age; that is, the age has an adverse effect on the residual lifetime. Negative ageing describes the opposite beneficial effect. If the same type of ageing pattern persists throughout the entire lifespan of a device, it is called monotonic ageing. However, in many practical situations, the effect of age does not remain same throughout the entire lifespan. Typically, the effect of age is initially beneficial where negative ageing takes place (the so-called burn-in phase). Next comes the 'useful life phase' where the ageing profile remains more or less the same and finally, the effect of age is adverse and the ageing is positive (the 'wear-out' phase). This kind of ageing is called nonmonotonic ageing and arises naturally in situations like infant mortality, work hardening of mechanical or electronic machines and lengths of political coalitions or divorce rates, etc.

To model this ageing phenomenon, several nonparametric ageing families of distributions have been defined in the literature. These classes have been categorized based on the behaviour of the survival function, failure rate function and the mean residual life (MRL) functions. The approach is to model lifetime as a nonnegative random variable X. For $t \geq 0$,

$$F(t) := P[X \leq t]$$

is called the *cumulative distribution function* (c.d.f.) of X and

$$\bar{F}(t) := P[X > t] = 1 - F(t)$$

denotes the corresponding *survival function*. For an absolutely continuous random variable X with *probability density function* (p.d.f.) $f(t)$, the failure rate function of the unit is defined by

$$r(t) := \frac{f(t)}{\bar{F}(t)}, \quad \text{for } t \geq 0 \text{ satisfying } \bar{F}(t) > 0.$$

It is easy to see that the failure rate function gives the approximate probability that a unit will fail in the interval $(t, t + \Delta t]$, given that the unit is functioning at time t. The MRL function answers the fundamental question 'what is the expected remaining life of a unit of age t?'

Definition 1.1 If F is a life distribution function with finite mean, then the MRL function of F at age $t > 0$ is defined as

$$e_F(t) := E(X - t|X > t) = \begin{cases} \frac{1}{\bar{F}(t)} \int_t^{\infty} \bar{F}(u)\mathrm{d}u & \text{if } \bar{F}(t) > 0 \\ 0 & \text{otherwise.} \end{cases} \quad (2)$$

We now recapitulate the definitions of some well-known ageing classes.

Definition 1.2 A life distribution F or survival function $\bar{F} := 1 - F$ having mean μ is said to be, or to have:

(i) increasing failure rate (IFR) if $\bar{F}(x+t)/\bar{F}(t)$ is decreasing in t whenever $x > 0$.
(ii) decreasing mean residual life (DMRL) if $\int_0^{\infty} \bar{F}(x+t)/\bar{F}(t)\mathrm{d}x$ is decreasing in t.
(iii) increasing failure rate average (IFRA) if $\left(\bar{F}(t)\right)^{\frac{1}{t}}$ is decreasing in $t > 0$.
(iv) new better than used (NBU) if $\bar{F}(x+t) \leq \bar{F}(x)\bar{F}(t)$ for all $x, t \geq 0$.
(v) new better than used in expectation (NBUE) if $\int_0^{\infty} \bar{F}(x+t)/\bar{F}(t)\mathrm{d}x \leq \int_0^{\infty} \bar{F}(x)\mathrm{d}x$.
(vi) harmonic new better than used in expectation (HNBUE) if $\int_t^{\infty} \bar{F}(x)\mathrm{d}x \leq \int_t^{\infty} \mathrm{e}^{-x/\mu}\mathrm{d}x$ for all $t \geq 0$.
(vii) new better than used failure rate (NBUFR) if $\frac{d}{dt} \ln \bar{F}(t) \leq \lim_{s \to 0^+} s^{-1} \ln \bar{F}(s)$ for all $t \geq 0$.
(viii) new better than average failure rate (NBAFR) if $t^{-1} \ln \bar{F}(t) \leq \lim_{s \to 0^+} s^{-1} \ln \bar{F}(s)$ for all $t \geq 0$.
(ix) $\mathscr{L}$ if $\int_0^{\infty} \mathrm{e}^{-st}\bar{F}(t)\mathrm{d}t \geq \mu/(1+s\mu)$ for all $s \geq 0$.

By reversing the direction of monotonicity or inequality in definitions (i)–(ix), the corresponding set of dual classes of distributions DFR, IMRL, DFRA, NWU, NWUE, HNWUE, NWUFR, NWAFR and $\bar{\mathscr{L}}$ are obtained, respectively. Here D = decreasing, I = increasing and W = worse. If F has a density, then F is IFR (DFR) if $r(x)$ is increasing (decreasing) in x. F is IFRA (DFRA) if $\frac{1}{x}\int_0^x r(u)du$ is increasing (decreasing) in x. F is NBUFR (NWUFR) if $r(0) \leq (\geq) r(x)$, for all $x > 0$ and NBAFR (NWAFR) if $r(0) \leq (\geq) \frac{1}{x}\int_0^x r(u)\mathrm{d}u$, for all $x > 0$. For details of the ageing classes discussed, the interested reader is referred to Bryson and Siddiqui (1969), Barlow and Proschan (1975), Rolski (1975), Klefsjö (1983), Hollander and Proschan (1984), Loh (1984), Abouammoh and Ahmed (1988), among others.

Among the above classes of life distributions prevalent in reliability analysis, we have the following well-known hierarchical relationships:

$$\begin{array}{ccccccccc} IFR & \Longrightarrow & IFRA & \Longrightarrow & NBU & \Longrightarrow & NBUFR & \Longrightarrow & NBAFR \\ \Downarrow & & & & \Downarrow & & & & \\ DMRL & & \Longrightarrow & & NBUE & \Longrightarrow & HNBUE & \Longrightarrow & \mathscr{L} \end{array}$$

For modelling of nonmonotonic ageing phenomena, one typically uses the failure rate and MRL functions. If failure rate function is used to model nonmonotonic ageing, then one can get the *bathtub failure rate* (BFR) class (see Glaser 1980) and when MRL function is employed, then we get the *increasing then decreasing mean residual life* (IDMRL) and *new worse then better than used in expectation* (NWBUE) ageing classes introduced by Guess et al. (1986) and Mitra and Basu (1994), respectively. It is worth noting that the NWBUE family is a rich class of life distributions encompassing both the IDMRL class as well as all BFR distributions. We now recapitulate the definitions of the above-mentioned nonmonotonic ageing classes.

Definition 1.3 A life distribution F is called a *bathtub failure rate* (BFR) (*upside-down bathtub failure rate* (UBFR)) distribution if there exists a point $t_0 \geq 0$ such that $-\ln \bar{F}(t)$ is concave (convex) on $[0, t_0)$ and convex (concave) on $[t_0, \infty)$. The point t_0 is referred to as a *change point* (or *turning point*) of the distribution in the BFR (UBFR) sense.

Definition 1.4 A life distribution F with a finite first moment is called an *increasing then decreasing mean residual life* (IDMRL) (*decreasing then increasing mean residual life* (DIMRL)) distribution if there exists a point $\tau \geq 0$ such that $e_F(t)$ is increasing (decreasing) on $[0, \tau)$ and decreasing (increasing) on $[\tau, \infty)$ where $e_F(t)$ is defined as in (2). The point τ is referred to as a *change point* (or *turning point*) of the distribution in the IDMRL (DIMRL) sense.

Definition 1.5 A life distribution F with a finite first moment is called an *new worse then better than used in expectation* (NWBUE) (*new better then worse than used in expectation* (NBWUE)) distribution if there exists a point $x_0 \geq 0$ such that $e_F(t) \geq e_F(0)$ for $[0, x_0)$ and $e_F(t) \leq e_F(0)$ for $[x_0, \infty)$ where $e_F(t)$ is defined in (2). The point x_0 is referred to as a *change point* (or *turning point*) of the distribution in the NWBUE (NBWUE) sense.

The results of Mitra and Basu (1994) and Khan et al. (2021) establish the following interrelationships between the BFR, IDMRL and NWBUE classes of life distributions:

$$\text{BFR} \Longrightarrow \text{IDMRL} \Longrightarrow \text{NWBUE}.$$

Several probabilistic and inferential aspects of the above mentioned ageing classes have received considerable attention in reliability literature. The properties of interest concerning ageing classes mainly involve preservation of class property under reliability operations like formation of coherent systems, convolutions and mixtures as well as issues regarding reliability bounds, moment bounds and moment inequalities, maintenance and replacement policies, closure under weak convergence, etc. For an extensive discussion on the aforementioned properties and their applications, one may refer to the outstanding books by Barlow and Proschan (1965, 1975), Zacks (1991), Lai and Xie (2006), Marshall and Olkin (2007) and others. In this article, we discuss how the above-mentioned nonparametric ageing classes arise from shock models.

The present paper is organized as follows. In Sect. 2, we introduce the definitions of the discrete versions of all the above mentioned ageing classes. Section 3 presents many of the more important results in shock model literature. A variety of scenarios where the arrival process ranges from homogeneous Poisson process (HPP), nonhomogeneous Poisson process (NHPP), pure birth shock model, etc., are considered. Finally in Sect. 4, we study the cumulative damage shock model.

2 The Discrete Definitions

An important aspect of reliability analysis is to find a life distribution that can adequately describe the ageing behaviour of the concerned device. Most of the lifetimes are continuous in nature and hence several continuous ageing classes have been proposed in the literature. On the other hand, discrete failure data where life is measured in terms of cycles, completed hours or completed number of shifts, etc., may arise in several common situations; for example, report on field failure is collected weekly, monthly and the observations are the number of failures, without specification of the failure time. The following definitions represent discrete ageing classes.

Definition 2.1 A survival probability $\bar{P}_k$ with support on $\{0, 1, 2, \ldots\}$ and with frequency function $p_k = \bar{P}_{k-1} - \bar{P}_k$ for $k = 1, 2, \ldots$, and $p_0 = 1 - \bar{P}_0$ is said to be or to have:

(i) discrete IFR (DFR) class if $\bar{P}_k/\bar{P}_{k-1}$ is decreasing (increasing) in $k = 1, 2, \ldots$.
(ii) discrete DMRL (IMRL) class if $\sum_{j=k}^{\infty} \bar{P}_j/\bar{P}_k$ is decreasing (increasing) in $k = 0, 1, 2, \ldots$.
(iii) discrete IFRA (DFRA) class if $\bar{P}_k^{1/k}$ is decreasing (increasing) in $k = 1, 2, \ldots$.
(iv) discrete NBU (NWU) class if $\bar{P}_j\bar{P}_k \geq (\leq)\bar{P}_{j+k}$ for $j, k = 0, 1, 2, \ldots$.
(v) discrete NBUE (NWUE) class if $\bar{P}_k \sum_{j=0}^{\infty} \bar{P}_j \geq (\leq) \sum_{j=k}^{\infty} \bar{P}_j$ for $k = 0, 1, 2, \ldots$.
(vi) discrete HNBUE (HNWUE) class if $\sum_{k=j}^{\infty} \bar{P}_k \leq (\geq) \sum_{k=0}^{\infty} \bar{P}_k \left(1 - \frac{1}{\sum_{k=0}^{\infty} \bar{P}_k}\right)^j$ for $j = 0, 1, 2, \ldots$.
(vii) discrete NBUFR (NWUFR) property if $\bar{P}_{k+1} \leq (\geq)\bar{P}_1\bar{P}_k$ for $k = 0, 1, 2, \ldots$.
(viii) discrete NBAFR (NWAFR) if $\bar{P}_k \leq (\geq)\bar{P}_1^{-k}$ for $k = 0, 1, 2, \ldots$.
(ix) discrete $\mathscr{L}$ ($\bar{\mathscr{L}}$)class if

$$\sum_{k=0}^{\infty} \bar{P}_k p^k \geq (\leq) \frac{\sum_{k=0}^{\infty} \bar{P}_k}{p + (1-p)\sum_{k=0}^{\infty} \bar{P}_k} \quad \text{for } 0 \leq p \leq 1.$$

These definitions can be found in Esary et al. (1973), Klefsjö (1981, 1983), Abouammoh and Ahmed (1988).

In the context of nonmonotonic ageing, the first results in this direction were obtained by Mitra and Basu (1996). They introduced the notions of discrete BFR and NWBUE distributions. In a similar vein, Anis (2012) provided the discrete version of the IDMRL ageing class. The discrete versions of these classes can formally be defined as follows:

Definition 2.2 A sequence $(\bar{P}_k)_{k=0}^{\infty}$ is said to possess the discrete BFR property if there exists a positive integer k_0 such that the following holds:

$$\frac{\bar{P}_1}{\bar{P}_0} \le \frac{\bar{P}_2}{\bar{P}_1} \le \frac{\bar{P}_3}{\bar{P}_2} \le \cdots \le \frac{\bar{P}_{k_0}}{\bar{P}_{k_0-1}} \ge \frac{\bar{P}_{k_0+1}}{\bar{P}_{k_0}} \ge \cdots$$

We say that k_0 is the change point of the sequence $\bar{P}_k$ (in the BFR sense) and write $(\bar{P}_k)_{k=0}^{\infty}$ is BFR(k_0).

Definition 2.3 A sequence $(\bar{P}_k)_{k=0}^{\infty}$ is said to possess the discrete IDMRL (DIMRL) property if there exists an integer $k_0 \ge 0$ such that

$$\frac{1}{\bar{P}_k}\sum_{j=k}^{\infty}\bar{P}_j \begin{cases} \text{is increasing (decreasing) } in\ k, & k < k_0, \\ \text{is decreasing (increasing) } in\ k, & k \ge k_0. \end{cases}$$

The point k_0 will be referred to as the change point of the sequence $(\bar{P}_k)_{k=0}^{\infty}$ (in the IDMRL (DIMRL) sense) and we shall write $(\bar{P}_k)_{k=0}^{\infty}$ is IDMRL(k_0) (DIMRL(k_0)).

Definition 2.4 A sequence $(\bar{P}_k)_{k=0}^{\infty}$ with $\sum_{j=0}^{\infty}\bar{P}_j < \infty$ is said to be a discrete NWBUE (NWBUE) sequence if there exists an integer $k_0 \ge 0$ such that

$$\bar{P}_k\sum_{j=0}^{\infty}\bar{P}_j - \sum_{j=k}^{\infty}\bar{P}_j \begin{cases} \le (\ge)\ 0 & \forall\ k < k_0, \\ \ge (\le)\ 0 & \forall\ k \ge k_0. \end{cases}$$

The point k_0 will be referred to as the change point of the sequence $(\bar{P}_k)_{k=0}^{\infty}$ (in the NWBUE (NBWUE) sense) and we shall write $(\bar{P}_k)_{k=0}^{\infty}$ is NWBUE(k_0) (NBWUE(k_0)).

3 The Shock Models

Reliability shock models are primarily concerned with the following fundamental question: if the sequence $(\bar{P}_k)_{k=0}^{\infty}$ possesses a certain discrete ageing property (such as IFR, IFRA, NBU, NBUE, etc.), does $\bar{H}(t)$, defined via the transformation (1), inherit the corresponding continuous version of the said ageing property in various scenarios (i.e. when $N(t)$ is a HPP, NHPP, stationary, or nonstationary pure birth process, etc.)?

3.1 Homogeneous Poisson Shock Model

One of the simplest models is the homogeneous Poisson shock model studied by Esary et al. (1973) in which the shocks occur according to a homogeneous Poisson process with constant intensity $\lambda > 0$, i.e.,

$$P(N(t) = k) = \mathrm{e}^{-\lambda t}\frac{(\lambda t)^k}{k!}, \quad k = 0, 1, \ldots.$$

The Poisson shock model has several applications in risk, survival, or reliability analysis, according as shocks arriving at the system are interpreted as the claims, the cause of deterioration of health, or the technical reasons for failure in a car or other machines during functioning. Therefore, it is interesting to study conditions on the discrete survival probabilities which give rise to the ageing classes under consideration.

Suppose that the device has probability $\bar{P}_k$ of surviving the first k shocks, $k = 0, 1, \ldots$. Then, the survival function of the device is given by

$$\bar{H}(t) = \sum_{k=0}^{\infty} \mathrm{e}^{-\lambda t}\frac{(\lambda t)^k}{k!}\bar{P}_k. \tag{3}$$

Esary et al. (1973) established that $\bar{H}(t)$ inherits the IFR, IFRA, NBU, NBUE and DMRL properties if the sequence $(\bar{P}_k)_{k=0}^{\infty}$ possesses the corresponding discrete properties. Analogous results for the HNBUE and $\mathscr{L}$ classes were proved in Klefsjö (1981, 1983), respectively. Later, Abouammoh et al. (1988) established such results for the NBUFR and NBAFR classes of life distributions.

Theorem 3.1 *Suppose that*

$$\bar{H}(t) = \sum_{k=0}^{\infty} e^{-\lambda t}\frac{(\lambda t)^k}{k!}\bar{P}_k,$$

where $1 = \bar{P}_0 \geq \bar{P}_1 \geq \bar{P}_2 \geq \cdots$. *Then,* H *belongs to* $\mathscr{C}$ *if* $(\bar{P}_k)_{k=0}^{\infty}$ *belongs to the discrete* $\mathscr{C}$ *class, where* $\mathscr{C}$ *is any one of the ageing classes such as IFR, DMRL, IFRA, NBU, NBUE, HNBUE, NBUFR, NBAFR,* $\mathscr{L}$ *or their respective duals.*

The key tools used to derive most of the above results is the notion of total positivity and in particular the variation diminishing property (VDP) of totally positive (TP) functions. In this context, we first recapitulate the definition of a totally positive (TP) function and a important theorem of Karlin (1968, p. 21).

Definition 3.1 Let $A, B \subseteq \mathbb{R}$. A function $K : A \times B \to \mathbb{R}$ is said to be totally positive of order n (we write K is TP_n) if for every $r = 1, 2, \ldots, n$,

$$\begin{vmatrix} K(x_1, y_1) & K(x_1, y_2) & \dots & K(x_1, y_r) \\ K(x_2, y_1) & K(x_2, y_2) & \dots & K(x_2, y_r) \\ \vdots & \vdots & \ddots & \vdots \\ K(x_r, y_1) & K(x_r, y_2) & \dots & K(x_r, y_r) \end{vmatrix} \geq 0$$

where $x_1 < x_2 < \cdots < x_r$ $(x_i \in A,\ i = 1, 2, \ldots, r)$ and $y_1 < y_2 < \cdots < y_r$ $(y_i \in B,\ i = 1, 2, \ldots, r)$.

Definition 3.2 A function which is TP_n for every $n \geq 1$ is called totally positive.

Consider a sequence of real numbers $x_1, x_2, \ldots x_m$ and $S(x_1, x_2, \ldots x_m)$ = no. of sign changes in $x_1, x_2, \ldots x_m$ disregarding zero terms. Let I be an ordered subset of $\mathbb{R}$ and f be a real-valued function defined on I. Define

$$S(f) := \sup_{t_1 < t_2 \cdots t_m \in I} S\left(f(t_1), f(t_2), \ldots, f(t_m)\right).$$

Theorem 3.2 (Karlin 1968, p. 21) *Suppose $K(x, y)$ is TP_r on $A \times B$ and f is a bounded, Borel measurable function defined on B. Let*

$$g(x) = \int_B K(x, y) f(y) dy.$$

Then,

(i) $S(g) \leq S(f)$ provided $S(f) \leq r - 1$.
(ii) f and g exhibit the same sequence of signs in the same direction.

To prove Theorem 3.1 in the context of IFR, IFRA, DMRL and their duals, we use the VDP of totally positive kernel $K(r, t) = \mathrm{e}^{-\lambda t} \dfrac{(\lambda t)^r}{r!}$, $r = \{0, 1, \ldots\}$, $t \in (0, \infty)$. Here, we will present the proof of Theorem 3.1 when $\mathscr{C}$ represents the IFRA class of life distributions. Analogously, using suitable modifications, one can prove this theorem for IFR and DMRL classes and their respective duals.

To prove the main result, we first need the following proposition.

Proposition 3.1 *If for each $\lambda > 0$, $\bar{F}(t) - e^{-\lambda t}$ has at most one change of sign from $+$ to $-$, then F is IFRA.*

Proof Suppose $0 < t_1 < t_2$. We want to show that $-\frac{\ln \bar{F}(t_1)}{t_1} \leq -\frac{\ln \bar{F}(t_2)}{t_2}$. Define $g(x) = -\ln \bar{F}(x)$ and $\lambda_i = \frac{g(t_i)}{t_i}$, $i = 1, 2$. Now, $\bar{F}(t) - \mathrm{e}^{-\lambda t}$ has at most one change of sign from $+$ to $-$ for each $\lambda > 0$. So this happens for λ_1 and λ_2 in particular. Thus, $g(x) - \lambda_i x$ has at most one change of sign from $-$ to $+$. By choice of λ_i, $i = 1, 2$, we have,

$$g(x) \begin{cases} \leq \lambda_i x \text{ for } x \leq t_i \\ \geq \lambda_i x \text{ for } x > t_i. \end{cases}$$

If possible, let $\lambda_1 = \frac{g(t_1)}{t_1} > \frac{g(t_2)}{t_2} = \lambda_2$. Take $t_1 < x < t_2$. Then

$$\begin{aligned} g(x) &\geq \lambda_1 x, \quad \text{as} \quad x > t_1 \\ &> \lambda_1 x, \quad \text{as} \quad \lambda_1 > \lambda_2 \\ &\geq g(x), \quad \text{as} \quad x < t_2. \end{aligned}$$

Thus, $g(x) > g(x)$, a contradiction. So $\frac{g(t_1)}{t_1} \leq \frac{g(t_2)}{t_2}$, i.e. $-\frac{\ln \bar{F}(t_1)}{t_1} \leq -\frac{\ln \bar{F}(t_2)}{t_2}$ and hence the theorem holds. □

Proof of Theorem 3.1 for the IFRA class

Consider $\eta \in [0, 1]$. As $\bar{P}_k^{1/k}$ is decreasing in k, $\bar{P}_k^{1/k} - \eta$ has at most one change of sign from + to −. (Note that $\bar{P}_0 = 1$ and furthermore observe that the graph of $\bar{P}_k^{1/k}$ will either stay above η or if it crosses η, it will subsequently remain below η as $\bar{P}_k^{1/k}$ is decreasing in k.) Recall that $e^{-\lambda t}\frac{(\lambda t)^k}{k!}$ is TP, $0 < t < \infty, k \in \{0, 1, 2, \ldots, \}$. Now

$$\begin{aligned} \bar{H}(t) - e^{-(1-\eta)\lambda t} &= \sum_{k=0}^{\infty} e^{-\lambda t}\frac{(\lambda t)^k}{k!}\bar{P}_k - e^{-\lambda t}\sum_{k=0}^{\infty}\frac{(\eta\lambda t)^k}{k!} \\ &= \sum_{k=0}^{\infty}\left(\bar{P}_k - \eta^k\right) e^{-\lambda t}\frac{(\lambda t)^k}{k!} \end{aligned}$$

Then, by the variation diminishing property (Theorem 3.2), $\bar{H}(t) - e^{-(1-\eta)\lambda t}$ has at most one change of sign from + to −. Take any $\theta > 0$. Clearly, there exists $\eta \in [0, 1]$ and some $\lambda > 0$ such that $\theta = (1 - \eta)\lambda$. Since θ is an arbitrary positive number, we can conclude that H is IFRA by virtue of Proposition 3.1. □

Now we will present the proof of Theorem 3.1 when $\mathscr{C}$ represents the NBU class of life distributions. One can prove (with some modifications) the theorem for the remaining above-mentioned monotonic ageing classes (see Esary et al. (1973), Klefsjö (1981, 1983), Abouammoh et al. (1988) among others).

Proof of Theorem 3.1 for the NBU class

$$\begin{aligned} \bar{H}(t)\bar{H}(x) &= \left(\sum_{k=0}^{\infty} e^{-\lambda t}\frac{(\lambda t)^k}{k!}\bar{P}_k\right)\left(\sum_{l=0}^{\infty} e^{-\lambda x}\frac{(\lambda x)^l}{l!}\bar{P}_l\right) \\ &= e^{-\lambda(t+x)}\left(\sum_{k=0}^{\infty}\frac{(\lambda t)^k}{k!}\bar{P}_k\right)\left(\sum_{l=0}^{\infty}\frac{(\lambda x)^l}{l!}\bar{P}_l\right) \\ &= e^{-\lambda(t+x)}\sum_{k=0}^{\infty}\sum_{j=0}^{k}\frac{(\lambda t)^{k-j}}{(k-j)!}\bar{P}_{k-j}\frac{(\lambda x)^j}{j!}\bar{P}_j \quad \left[\text{as } \left(\sum_{k=0}^{\infty} a_k \sum_{k=0}^{\infty} b_k\right) = \sum_{k=0}^{\infty} c_k \text{ where } c_k = \sum_{j=0}^{k} a_{k-j}b_j\right] \\ &\geq e^{-\lambda(t+x)}\sum_{k=0}^{\infty}\sum_{j=0}^{k}\frac{k!}{j!(k-j)!k!}(\lambda t)^{k-j}(\lambda x)^j\bar{P}_k \quad \left[\text{as } \bar{P}_{k-j}\bar{P}_j \geq \bar{P}_k\right] \end{aligned}$$

$$= e^{-\lambda(t+x)} \sum_{k=0}^{\infty} \frac{\bar{P}_k}{k!} \sum_{j=0}^{k} \infty \binom{k}{j} (\lambda t)^{k-j} (\lambda x)^j$$

$$= e^{-\lambda(t+x)} \sum_{k=0}^{\infty} \frac{\bar{P}_k}{k!} (\lambda x + \lambda t)^k = \sum_{k=0}^{\infty} e^{-\lambda(t+x)} \frac{(\lambda(t+x))^k}{k!} \bar{P}_k = \bar{H}(t+x)$$

Thus, $\bar{H}(t)\bar{H}(x) \geq \bar{H}(t+x)$ and hence H in NBU. □

For the nonmonotonic ageing classes, the corresponding results available in the literature are as follows.

Theorem 3.3 *Let $(\bar{P}_k)_{k=0}^{\infty}$ be BFR(k_0), $k_0 > 0$ and $\bar{H}(t)$ is defined as in (3). Suppose that*

$$\frac{\bar{P}_{k_0+1}}{\bar{P}_{k_0}} \leq \frac{\bar{P}_1}{\bar{P}_0}$$

holds. If

(i) $k_0 \leq 3$, then $\bar{H}$ is BFR;
(ii) $k_0 > 3$ and $b_r := \sum_{j=0}^{r} (\bar{P}_{j+2}\bar{P}_{r-j} - \bar{P}_{j+1}\bar{P}_{r-j+1})/j!(r-j)!$ has the same sign for all r, $k_0 - 1 \leq r \leq 2k_0 - 4$, then $\bar{H}$ is BFR.

Theorem 3.4 *Consider a discrete NWBUE(k_0) (NBWUE(k_0)) sequence $(\bar{P}_k)_{k=0}^{\infty}$ and let $\bar{H}(t)$ be defined as in (3). Then, H is NWBUE (NBWUE).*

For details of proofs, one can see the first paper in this direction by Mitra and Basu (1996). To prove the corresponding result for IDMRL (DIMRL) distributions, the following lemma due to Belzunce et al. (2007) is crucial.

Lemma 3.5 (Belzunce et al. 2007) *Suppose that $\{\alpha_k/\beta_k\}_{k=0}^{\infty}$ is a real sequence with the property that*

$$\frac{\alpha_k}{\beta_k} \begin{cases} \text{is increasing (decreasing) in } k \text{ for } k < k_0 \\ \text{is decreasing (increasing) in } k \text{ for } k \geq k_0 \end{cases}$$

for some positive integer k_0 and also assume that the power series defined by $\phi(s) := \sum_{k=0}^{\infty} \frac{\alpha_k}{k!} s^k$, $s \geq 0$ and $\psi(s) := \sum_{k=0}^{\infty} \frac{\beta_k}{k!} s^k$, $s \geq 0$ converge absolutely $\forall$ $s \geq 0$. Then, $\exists$ $s_0 \in \mathbb{R}$ such that

$$\frac{\phi(s)}{\psi(s)} \begin{cases} \text{is increasing (decreasing) in } s \text{ for } s < s_0 \\ \text{is decreasing (increasing) in } s \text{ for } s \geq s_0. \end{cases}$$

Theorem 3.6 *Consider a discrete IDMRL(k_0) (DIMRL(k_0)) sequence $(\bar{P}_k)_{k=0}^{\infty}$ and let $\bar{H}(t)$ be defined as in (3). Then, H is IDMRL (DIMRL).*

Proof Note that

$$\int_t^{\infty} \bar{H}(u)\mathrm{d}u = \int_t^{\infty} \sum_{k=0}^{\infty} \mathrm{e}^{-\lambda u} \frac{(\lambda u)^k}{k!} \bar{P}_k \mathrm{d}u$$

$$= \sum_{k=0}^{\infty} \frac{\bar{P}_k}{\lambda} \int_t^{\infty} \frac{\lambda^{k+1}}{\Gamma(k+1)} \mathrm{e}^{-\lambda u} u^k \mathrm{d}u$$

$$= \sum_{k=0}^{\infty} \sum_{i=0}^{k} \frac{\bar{P}_k}{\lambda} \frac{(\lambda t)^i}{i!} \mathrm{e}^{-\lambda t}$$

$$= \frac{\mathrm{e}^{-\lambda t}}{\lambda} \sum_{i=0}^{\infty} \frac{(\lambda t)^i}{i!} \left(\sum_{k=i}^{\infty} \bar{P}_k \right).$$

Now

$$e_H(t) = \frac{\frac{\mathrm{e}^{-\lambda t}}{\lambda} \sum_{i=0}^{\infty} \frac{(\lambda t)^i}{i!} \left(\sum_{k=i}^{\infty} \bar{P}_k \right)}{\sum_{k=0}^{\infty} \mathrm{e}^{-\lambda t} \frac{(\lambda t)^k}{k!} \bar{P}_k} = \frac{1}{\lambda} \frac{\sum_{k=0}^{\infty} \frac{(\lambda t)^k}{k!} \sum_{j=k}^{\infty} \bar{P}_j}{\sum_{k=0}^{\infty} \frac{(\lambda t)^k}{k!} \bar{P}_k} = \frac{1}{\lambda} \delta(t)$$

where

$$\delta(t) = \frac{\sum_{k=0}^{\infty} \frac{(\lambda t)^k}{k!} \sum_{j=k}^{\infty} \bar{P}_j}{\sum_{k=0}^{\infty} \frac{(\lambda t)^k}{k!} \bar{P}_k}.$$

In order to prove the theorem, we simply need to show that there exists a $t_0 \geq 0$ for which

$$\delta(t) = \begin{cases} \text{is increasing in } t \text{ for } t < t_0 \\ \text{is decreasing in } t \text{ for } t \geq t_0. \end{cases} \tag{4}$$

Now define $\alpha_k = \sum_{j=k}^{\infty} \bar{P}_j$ and $\beta_k = \bar{P}_k$ in $\delta(t)$. Then,

$$\frac{\alpha_k}{\beta_k} = \frac{1}{\bar{P}_k} \sum_{j=k}^{\infty} \bar{P}_j \begin{cases} \text{is increasing in } k \text{ for } k < k_0 \\ \text{is decreasing in } k \text{ for } k \geq k_0 \end{cases} \tag{5}$$

as $(\bar{P}_k)_{k=0}^{\infty}$ has the discrete IDMRL property. Putting $s = \lambda t$, an application of Lemma 3.5 yields

$$\delta(t) = \frac{\sum_{k=0}^{\infty} \frac{(\lambda t)^k}{k!} \sum_{j=k}^{\infty} \bar{P}_j}{\sum_{k=0}^{\infty} \frac{(\lambda t)^k}{k!} \bar{P}_k} = \frac{\phi(s)}{\psi(s)} \begin{cases} \text{is increasing (decreasing) in } s \text{ for } s < s_0 \\ \text{is decreasing (increasing) in } s \text{ for } s \geq s_0. \end{cases}$$

Thus H is IDMRL. □

3.2 Nonhomogeneous Poisson Shock Model

This shock model was first studied by A-Hameed and Proschan (1973). They extended the work of Esary et al. (1973) to the context of nonhomogeneous Poisson process. In this model, shocks arrive according to a nonhomogeneous Poisson process with intensity function $\Lambda(t)$ and event rate $\lambda(t) = \mathrm{d}\Lambda(t)/\mathrm{d}t$, both defined on the domain $[0, \infty)$; we take $\lambda(0)$ as the right-hand derivative of $\Lambda(t)$ at $t = 0$. In this case, the survival function $\bar{H}(t)$ can be expressed as

$$\bar{H}^*(t) = \sum_{k=0}^{\infty} \mathrm{e}^{-\Lambda(t)} \frac{[\Lambda(t)]^k}{k!} \bar{P}_k, \tag{6}$$

with density

$$h^*(t) = \sum_{k=0}^{\infty} \mathrm{e}^{-\Lambda(t)} \frac{[\Lambda(t)]^k}{k!} \lambda(t) p_{k+1}.$$

With suitable assumptions on $\Lambda(t)$, A-Hameed and Proschan (1973) established that if $\left(\bar{P}_k\right)_{k=0}^{\infty}$ has the discrete IFR, IFRA, NBU, NBUE or DMRL property, then the survival function $\bar{H}(t)$ has the corresponding continuous property. Klefsjö (1981, 1983) established the same result for the HNBUE and $\mathscr{L}$ classes of life distributions. Later, analogous theorems for NBUFR and NBAFR families were proved by Abouammoh et al. (1988). The following theorem collects all the above-mentioned results.

Theorem 3.7 *Suppose that*

$$\bar{H}^*(t) = \sum_{k=0}^{\infty} e^{-\Lambda(t)} \frac{[\Lambda(t)]^k}{k!} \bar{P}_k$$

where $1 = \bar{P}_0 \geq \bar{P}_1 \geq \bar{P}_2 \geq \cdots$. *Then*

(i) $\bar{H}^*(t)$ *is IFR (DFR) if* $(\bar{P}_k)_{k=0}^{\infty}$ *belongs to the discrete IFR (DFR) class and* $\Lambda(t)$ *is convex (concave).*
(ii) $\bar{H}^*(t)$ *is IFRA (DFRA) if* $(\bar{P}_k)_{k=0}^{\infty}$ *belongs to the discrete IFRA (DFRA) class and* $\Lambda(t)$ *is starshaped (anti-starshaped).*
(iii) $\bar{H}^*(t)$ *is DMRL (IMRL) if* $(\bar{P}_k)_{k=0}^{\infty}$ *belongs to the discrete DMRL (IMRL) class and* $\Lambda(t)$ *is convex (concave).*
(iv) $\bar{H}^*(t)$ *is NBU (NWU) if* $(\bar{P}_k)_{k=0}^{\infty}$ *belongs to the discrete NBU (NWU) class and* $\Lambda(t)$ *is superadditive (subadditive).*
(v) $\bar{H}^*(t)$ *is NBUE (NWUE) if* $(\bar{P}_k)_{k=0}^{\infty}$ *belongs to the discrete NBUE (NWUE) class and* $\Lambda(t)$ *is starshaped (anti-starshaped).*
(vi) $\bar{H}^*(t)$ *is HNBUE (HNWUE) if* $(\bar{P}_k)_{k=0}^{\infty}$ *belongs to the discrete HNBUE (HNWUE) class and* $\Lambda(t)$ *is starshaped (anti-starshaped).*
(vii) $\bar{H}^*(t)$ *is NBUFR (NWUFR) if* $(\bar{P}_k)_{k=0}^{\infty}$ *belongs to the discrete NBUFR (NWUFR) class and* $\Lambda(0) = 0$ *and* $\Lambda^{'}(t) > \Lambda^{'}(0)$ $(\Lambda^{'}(t) < \Lambda^{'}(0))$.

(viii) $\bar{H}^*(t)$ *is NBAFR (NWUFR) if* $(\bar{P}_k)_{k=0}^{\infty}$ *belongs to the discrete NBUFR (NWUFR) class and* $\Lambda(0) = 0$ *and* $\Lambda(t) > t\Lambda^{'}(0)$ $(\Lambda(t) < t\Lambda^{'}(0))$.

(ix) $\bar{H}^*(t)$ *is* $\mathcal{L}$ $(\bar{\mathcal{L}})$ *if* $(\bar{P}_k)_{k=0}^{\infty}$ *belongs to the discrete* $\mathcal{L}$ $(\bar{\mathcal{L}})$ *class and* $\Lambda(t)$ *is starshaped (anti-starshaped).*

The proofs of the results contained in the above theorem utilize the following lemmas concerning composition of functions. Note that a nonnegative function $f(x)$ defined on $[0, \infty)$ is superadditive (subadditive) if $f(x+y) \geq (\leq) f(x) + f(y)$ for all $x \geq 0, y \geq 0$. A nonnegative function g is defined on $[0, \infty)$ with $g(0) = 0$ is said to be starshaped (anti-starshaped) if $\frac{g(x)}{x}$ is increasing (decreasing) on $(0, \infty)$.

Lemma 3.8 *Let* $u(t) = u_1(u_2(t))$ *and let* u_1 *be increasing. Then,*

(a) u_i *convex (concave),* $i = 1, 2$, $\implies$ u *convex (concave).*

(b) u_i *starshaped,* $i = 1, 2$, $\implies$ u *starshaped.*

(c) $u_i(\alpha x) \geq \alpha u_i(x)$ *for all* $0 \leq \alpha \leq 1$ *and* $x \geq 0$, $i = 1, 2$, $\implies$ $u(\alpha x) \geq \alpha u(x)$ *for all* $0 \leq \alpha \leq 1$ *and* $x \geq 0$.

(d) u_i *superadditive (subadditive),* $i = 1, 2$, $\implies$ u *superadditive (subadditive).*

Lemma 3.9 (a) *Let* F *be DMRL (IMRL) and* g *be an increasing and convex (concave) function. Then* $K(t) = F(g(t))$ *is DMRL (IMRL).*

(b) *Let* F *be NBUE (NWUE) and* g *be an increasing and starshaped function (*g *be an increasing function such that* $g(\alpha x) \geq \alpha g(x)$, $0 \leq \alpha \leq 1$*). Then* $K(t) = F(g(t))$ *is NBUE (NWUE).*

Note that $\bar{H}^*(t) = \bar{H}(\Lambda(t))$ where $\bar{H}(t)$ is defined as in (3). Now applying Theorem 3.1 and the above lemmas, one can prove Theorem 3.7 after imposing a suitable condition on $\Lambda(t)$.

3.3 Pure Birth Shock Model

Now consider a pure birth shock model where a system is subjected to shocks governed by a birth process with intensities $\lambda_k, k = 0, 1, 2, \ldots$. Let V_{k+1} denote the inter-arrival time between the k-th and $(k+1)$-st shocks. Assume that V_k's are independent with V_{k+1} being exponentially distributed with mean $1/\lambda_k$ for $k = 0, 1, 2, \ldots$. Then, the survival function $\bar{H}(t)$ of the system can be written as

$$\bar{H}(t) = \sum_{k=0}^{\infty} Z_k(t)\bar{P}_k \tag{7}$$

with

$$Z_k(t) := P\left(N(t) = k\right),$$

where $N(t)$ is the pure birth process defined above. We assume that the intensities $\{\lambda_k\}_{k=0}^{\infty}$ are such that the probability of infinitely many shocks in $(0, t]$ is 0 (i.e.

$\sum_{k=0}^{\infty} Z_k(t) = 1$), which is equivalent to the condition $\sum_{k=0}^{\infty} \lambda_k^{-1} = \infty$; see Feller (1968, p. 452).

Note that $Z_k(t)$ is TP since the intervals between successive shocks in a stationary pure birth process are independent (nonidentical) exponential random variables (see Karlin and Proschan (1960), Theorem 3). To develop sufficient conditions for the preservation of ageing properties, we will use the following lemmas.

Lemma 3.10 (A-Hameed and Proschan 1975) *Let $Z_k(t)$ be as defined in (7). Then*

$$(a)\quad \int_0^t Z_k(u)du = \frac{1}{\lambda_k}\sum_{j=k+1}^{\infty} Z_j(t),$$

$$(b)\quad \int_t^{\infty} Z_k(u)du = \frac{1}{\lambda_k}\sum_{j=0}^{k} Z_j(t).$$

With suitable assumptions on λ_k, A-Hameed and Proschan (1973) established that if $\left(\bar{P}_k\right)_{k=0}^{\infty}$ has the discrete IFR, IFRA, NBU, NBUE or DMRL property, then under some suitable assumptions on $\Lambda(t)$, the survival function $\bar{H}(t)$ has the corresponding continuous property. Klefsjö (1981, 1983) established analogous results for the HNBUE and $\mathscr{L}$ classes of life distributions respectively. Later, similar results for NBUFR and NBAFR families were proved by Abouammoh et al. (1988). The following theorem collects all the above-mentioned results.

Theorem 3.11 *Suppose that*

$$\bar{H}(t) = \sum_{k=0}^{\infty} Z_k(t)\bar{P}_k$$

where $1 = \bar{P}_0 \geq \bar{P}_1 \geq \bar{P}_2 \geq \ldots$. Then,

(i) $\bar{H}(t)$ is IFR if λ_k is increasing in $k = 0, 1, 2, \ldots$ and $\bar{P}_k$ is discrete IFR.
(ii) $\bar{H}(t)$ is IFRA if λ_k is increasing in $k = 0, 1, 2, \ldots$ and $\bar{P}_k$ is discrete IFRA.
(iii) $\bar{H}(t)$ is NBU if λ_k is increasing in $k = 0, 1, 2, \ldots$ and $\bar{P}_k$ is discrete NBU.
(iv) $\bar{H}(t)$ is DMRL if λ_k is increasing in $k = 0, 1, 2, \ldots$ and $\bar{P}_k^{-1}\sum_{j=k}^{\infty}(\bar{P}_j/\lambda_j)$ is decreasing in $k = 0, 1, 2, \ldots$.
(v) $\bar{H}(t)$ is NBUE if $\sum_{j=0}^{\infty}(\bar{P}_j/\lambda_j) \geq \bar{P}_k^{-1}\sum_{j=k}^{\infty}(\bar{P}_j/\lambda_j)$ for $k = 0, 1, 2, \ldots$.
(vi) $\bar{H}(t)$ is HNBUE if $\sum_{j=k}^{\infty}(\bar{P}_j/\lambda_j) \leq \alpha_0\bar{P}_k^{-1}\sum_{j=0}^{k-1}(1-(\alpha_0\lambda_j)^{-1})$ for $k = 1, 2,$ *…, where $\alpha_0 = \sum_{j=0}^{\infty}(\bar{P}_j/\lambda_j)$.*
(vii) $\bar{H}(t)$ is NBUFR if $\lambda_k \geq \lambda_0$ for $k = 0, 1, \ldots$ and $(\bar{P}_k)_{k=0}^{\infty}$ belongs to the discrete NBUFR class.
(viii) $\bar{H}(t)$ is NBAFR if $\lambda_k > \lambda_0$ for $k = 0, 1, \ldots$ and $(\bar{P}_k)_{k=0}^{\infty}$ belongs to the discrete NBAFR class.

(ix) $\bar{H}(t)$ *is* $\mathcal{L}$ *if*

$$\sum_{k=0}^{\infty} \bar{P}_k \pi_k(s) \geq \frac{\alpha_0}{1 + s\alpha_0} \text{ for } s \geq 0.$$

where $\alpha_0 = \sum_{j=0}^{\infty}(\bar{P}_j/\lambda_j)$ *and* $\pi_k(s)$ *is defined by*

$$\pi_0(s) = \frac{1}{\lambda_0 + s} \textit{ and } \pi_k(s) = \left(\prod_{j=0}^{k-1} \frac{\lambda_j}{\lambda_j + s}\right) \frac{1}{\lambda_k + s} \text{ for } k = 1, 2, \ldots.$$

Using the variation diminishing property of totally positive (TP) functions and Lemma 3.10, whenever required, one can prove the above theorem applying arguments analogous to those in the proof of Theorem 3.1.

In this subsection, we treat the more general case in which shocks occur according to the following nonstationary pure birth process: *shocks occur according to a Markov process; given that k shocks have occurred in* $(0, t]$, *the probability of a shock occurring in* $(t, t + \Delta t]$ *is* $\lambda_k \lambda(t) \Delta t + o(\Delta t)$, *while the probability of more than one shock occurring in* $(t, t + \Delta t)$ *is* $o(\Delta t)$.

Remark 3.12 Note that in the stationary pure birth process, given that k shocks have occurred in $[0, \Lambda(t)]$, the probability of a shock occurring in $[\Lambda(t), \Lambda(t) + \lambda(t)\Delta t]$ (where $\Lambda(t) = \int_0^{\infty} \lambda(x)dx$) is of the same form: $\lambda_k \lambda(t) \Delta t + o(\Delta t)$. is of the same form: $\lambda_k \lambda(t) \Delta + o(\Delta)$. It follows immediately that the pure birth shock model may be obtained from the stationary pure birth process by the transformation $t \to \Lambda(t)$.

This model was first studied by A-Hameed and Proschan (1975). By making appropriate assumptions on λ_k and $\lambda(t)$, they proved that if the survival probabilities $(\bar{P}_k)_{k=0}^{\infty}$ possess discrete IFR, IFRA, NBU, NBUE properties (or their duals), then the continuous time survival probability $\bar{H}(t)$ belongs respectively to IFR, IFRA, NBU, NBUE (or their dual) classes. Klefsjö (1981, 1983) proved that the same results remain valid for the HNBUE and $\mathscr{L}$-families as well. Later, Abouammoh et al. (1988) showed this for NBUFR and NBAFR classes of life distributions.

The survival function $\bar{H}^*(t)$ of the system for this shock model can be written as

$$\bar{H}^*(t) = \sum_{k=0}^{\infty} Z_k(\Lambda(t))\bar{P}_k, \tag{8}$$

where $Z_k(t)$ is as defined in (7) and $\Lambda(t) = \int_0^t \lambda(u)\mathrm{d}u$.

Theorem 3.13 *Suppose that*

$$\bar{H}^*(t) = \sum_{k=0}^{\infty} Z_k(\Lambda(t))\bar{P}_k$$

where $1 = \bar{P}_0 \geq \bar{P}_1 \geq \bar{P}_2 \geq \ldots.$ *Then,* *(i)* $\bar{H}^*(t)$ *is IFR if* λ_k *is increasing in* $k = 0, 1, 2, \ldots$*,* $\Lambda(t)$ *is convex and* $(\bar{P}_k)_{k=0}^{\infty}$ *belongs to the discrete IFR class.*

(ii) $\bar{H}^*(t)$ *is IFRA if* λ_k *is increasing in* $k = 0, 1, 2, \ldots$*,* $\Lambda(t)$ *be starshaped and* $(\bar{P}_k)_{k=0}^{\infty}$ *belongs to the discrete IFRA class.*

(iii) $\bar{H}^*(t)$ *is NBU if* λ_k *is increasing in* $k = 0, 1, 2, \ldots,$*,* $\Lambda(t)$ *be superadditive and* $(\bar{P}_k)_{k=0}^{\infty}$ *belongs to the discrete NBU class.*

(iv) $\bar{H}^*(t)$ *is DMRL if* λ_k *is increasing in* $k = 0, 1, 2, \ldots$ *and* $\bar{P}_k^{-1} \sum_{j=k}^{\infty} (\bar{P}_j/\lambda_j)$ *is decreasing in* $k = 0, 1, 2, \ldots.$*.*

(v) $\bar{H}^*(t)$ *is NBUE if* $\Lambda(t)$ *be starshaped and* $\sum_{j=0}^{\infty} (\bar{P}_j/\lambda_j) \geq \bar{P}_k^{-1} \sum_{j=k}^{\infty} (\bar{P}_j/\lambda_j)$ for $k = 0, 1, 2, \ldots.$

(vi) $\bar{H}^*(t)$ *is HNBUE if* $\Lambda(t)$ *be starshaped and* $\sum_{j=k}^{\infty} (\bar{P}_j/\lambda_j) \leq \alpha_0 \bar{P}_k^{-1} \sum_{j=0}^{k-1} (1 - (\alpha_0 \lambda_j)^{-1})$ for $k = 1, 2, \ldots$ *where* $\alpha_0 = \sum_{j=0}^{\infty} (\bar{P}_j/\lambda_j)$.

(vii) $\bar{H}^*(t)$ *is NBUFR if* $\lambda_k \geq \lambda_0$ *for* $k = 0, 1, \ldots$*,* $\lambda(t) \geq \lambda(0)$ *and* $(\bar{P}_k)_{k=0}^{\infty}$ *belongs to the discrete NBUFR class.*

(viii) $\bar{H}^*(t)$ *is NBAFR if* $\lambda_k > \lambda_0$ *for* $k = 0, 1, \ldots$*,* $\Lambda(t) > t\Lambda'(0)$ *and* $(\bar{P}_k)_{k=0}^{\infty}$ *belongs to the discrete NBAFR class.*

(ix) $\bar{H}^*(t)$ *is* $\mathscr{L}$ *if* $\Lambda(t)$ *is starshaped and*

$$\sum_{k=0}^{\infty} \bar{P}_k \pi_k(s) \geq \frac{\alpha_0}{1 + s\alpha_0} \quad \text{for } s \geq 0$$

where $\alpha_0 = \sum_{j=0}^{\infty} (\bar{P}_j/\lambda_j)$ *and* $\pi_k(s)$ *is defined by*

$$\pi_0(s) = \frac{1}{\lambda_0 + s} \quad \text{and} \quad \pi_k(s) = \left(\prod_{j=0}^{k-1} \frac{\lambda_j}{\lambda_j + s} \right) \frac{1}{\lambda_k + s} \quad \text{for } k = 1, 2, \ldots.$$

Note that $\bar{H}^*(t) = \bar{H}(\Lambda(t))$ where $\bar{H}(t)$ is defined as in (7). Now applying Theorem 3.11 together with Lemmas 3.8 and 3.9, one can prove Theorem 3.13 after imposing a suitable condition on $\Lambda(t)$.

4 The Cumulative Damage Shock Model

In this model, each arriving shock causes a damage of random magnitude to the device. When the accumulated damage exceeds a certain threshold x, the device fails.

Let X_i be the damage caused to the equipment by the i-th shock, $i = 1, 2, \ldots.$ We assume that $X_1, X_2, \ldots$ are independent and identically distributed (iid) with some common distribution function (df) F. We also assume that shocks arrive according to homogeneous Poisson process with intensity $\lambda > 0$. In this case, we shall denote the survival function of the system by $\bar{H}_F$ to indicate the dependence on F.

$$\begin{aligned}\bar{H}_F(t) &= P[\text{the device survives beyond time t}]\\ &= \sum_{k=0}^{\infty} P[X_1 + X_2 + \cdots + X_{N(t)} \le x \mid N(t) = k] P[N(t) = k]\\ &= \sum_{k=0}^{\infty} P[X_1 + X_2 + \cdots + X_k \le x] \mathrm{e}^{-\lambda t} \frac{(\lambda t)^k}{k!}\\ &= \sum_{k=0}^{\infty} \mathrm{e}^{-\lambda t} \frac{(\lambda t)^k}{k!} F^{[k]}(x) \qquad (9)\end{aligned}$$

where $F^{[k]}(x) = (F * F * \cdots F)(x)$ is the k-fold convolution of F with itself. We write $F^{[0]}(x) = 1$ or 0 according as $x \ge 0$ or $x < 0$. So, $F^{[k]}(x) = P[X_1 + X_2 + \ldots + X_k \le x] = P[\text{surviving } k \text{ shocks}] = \bar{P}_k$ according to our previously adopted notation.

Lemma 4.1 *For every df F with $F(x) = 0$ for $x < 0$, $\left(F^{[k]}(x)\right)^{1/k}$ decreasing in k.*

Proof The proof is by induction. $F^{[2]}(x) = \int_0^x F(x - y) \mathrm{d}F(y) \le F(x) \int_0^x dF(y) = (F(x))^2$ as $F(x - y) \le F(x)$ since F is nondecreasing. Thus, $\left(F^{[2]}(x)\right)^{1/2} \le F(x)$. Assume that

$$\left(F^{[k-1]}(x)\right)^{1/k-1} \ge \left(F^{[k]}(x)\right)^{1/k}, \quad k \ge 2. \qquad (10)$$

Now,

$$\begin{aligned}\left(F^{[k]}(x)\right)^{k+1} &= F^{[k]}(x) \left(F^{[k]}(x)\right)^k\\ &= F^{[k]}(x) \left(\int_0^x F^{[k-1]}(x - y) \mathrm{d}F(y)\right)^k\\ &\ge \left(\left\{F^{[k]}(x)\right\}^{\frac{1}{k}} \int_0^x \left\{F^{[k]}(x - y)\right\}^{\frac{k-1}{k}} \mathrm{d}F(y)\right)^k, \text{ by (10)}\\ &\ge \left(\int_0^x F^{[k]}(x - y) \mathrm{d}F(y)\right)^k, \text{ as } F^{[k]}(x) \ge F^{[k]}(x - y) \text{ since } F^{[k]} \text{ is a df}\\ &= \left(F^{[k+1]}(x)\right)^k.\end{aligned}$$

Raising both sides to the power $\frac{1}{k(k+1)}$, we get,

$$\left(F^{[k]}(x)\right)^{1/k} \ge \left(F^{[k+1]}(x)\right)^{1/k+1}, \qquad (11)$$

which completes the induction step and the proof. □

As $\bar{P}_k = F^{[k]}$ in the cumulative damage model, the lemma shows that the condition '$\bar{P}_k^{1/k}$ is decreasing in k' is satisfied here irrespective of the choice of F. We thus have

Theorem 4.2 *$\bar{H}_F$ is IFRA irrespective of the choice of F.*

Proof Trivial using the above lemma and Theorem 3.1. □

At this juncture, a couple of comments would be in order. The condition '$\bar{P}_k^{1/k}$ is decreasing in k' in Theorem 3.1 can be described as 'the sequence $\bar{P}_k$ has the discrete IFRA property' (recall that F is IFRA is equivalent to $\left(\bar{F}(t)\right)^{1/t}$ is decreasing in t). Thus, IFRA distributions occur very naturally through such cumulative damage models; observe that there is no obvious IFRA property which is being inherited.

Now, we will discuss a cumulative damage shock model with independent but non-identically distributed random damages. The situation is identical to the previous model except that the X_i's are not identically distributed. So, $X_1, X_2, \ldots$ are damages that are independent and let $F_k(x) := P(X_k \leq x)$. Assume $F_k(x)$ is decreasing in k for each $x \geq 0$, i.e.

$$F_1(x) \geq F_2(x) \geq F_3(x) \cdots \quad \text{for each } x \geq 0. \tag{12}$$

The assumption that $F_k(x)$ is decreasing in k for each x has the interpretation that shocks are increasingly more effective in causing damage/wear to the device. Here

$$\bar{H}(t) = \sum_{k=0}^{\infty} \mathrm{e}^{-\lambda t} \frac{(\lambda t)^k}{k!} (F_1 * F_2 * \cdots * F_k)(x) \tag{13}$$

Observe that in this case,

$$\bar{P}_k = P[X_1 + \cdots + X_k \leq x] = (F_1 * F_2 * \cdots * F_k)(x) \tag{14}$$

Lemma 4.3 *If $F_i(x) = 0$ for $x < 0$ for all $i = 1, 2, \ldots$, then $[(F_1 * F_2 * \cdots * F_k)]^{\frac{1}{k}}$ is decreasing in k for each $x \geq 0$.*

Proof Using mathematical induction, we prove this lemma. Note that

$$\begin{aligned}
(F_1 * F_2)(x) &= \int_0^x F_2(x-y)\mathrm{d}F_1(y), \ \text{as} F_1 * F_2 = F_2 * F_1 \\
&\leq F_2(x) \int_0^x \mathrm{d}F_1(y), \ \text{as } F_2(x) \geq F_2(x-y) \\
&\leq (F_1(x))^2, \ \text{by (12)}.
\end{aligned}$$

Hence $[(F_1 * F_2)(x)]^{1/2} \leq F_1(x)$. Assume that

$$\left[(F_1 * F_2 * \cdots * F_{k-1})\right]^{\frac{1}{k-1}} \geq [(F_1 * F_2 * \cdots * F_k)]^{\frac{1}{k}}, \quad k \geq 2. \tag{15}$$

Now

$$\begin{aligned}
[(F_1 * F_2 * \cdots * F_k)(x)]^{k+1} &= (F_1 * F_2 * \cdots * F_k)(x)\,[(F_1 * F_2 * \cdots * F_k)(x)]^k \\
&= (F_1 * F_2 * \cdots * F_k)(x) \left[\int_0^x (F_1 * F_2 * \cdots * F_{k-1})(x-y)\mathrm{d}F_k(y)\right]^k \\
&\geq (F_1 * F_2 * \cdots * F_k)(x) \left[\int_0^x \{(F_1 * F_2 * \cdots * F_k)(x-y)\}^{\frac{k-1}{k}}\, dF_k(y)\right]^k \\
&= \Big[\{(F_1 * F_2 * \cdots * F_k)(x)\}^{\frac{1}{k}} \times \\
&\qquad \int_0^x \{(F_1 * F_2 * \cdots * F_k)(x-y)\}^{\frac{k-1}{k}}\, \mathrm{d}F_k(y)\Big]^k, \quad \text{by (15)} \\
&\geq \left[\int_0^x (F_1 * F_2 * \cdots * F_k)(x-y)\mathrm{d}F_k(y)\right]^k, \\
&\qquad \text{as} (F_1 * F_2 * \cdots * F_k)(x) \geq (F_1 * F_2 * \cdots * F_k)(x-y) \\
&\geq \left[\int_0^x (F_1 * F_2 * \cdots * F_k)(x-y)\mathrm{d}F_{k+1}(y)\right]^k \quad \text{as } F_k(y) \geq F_{k+1}(y) \forall\, y \geq 0 \\
&\geq \left[(F_1 * F_2 * \cdots * F_{k+1})(x)\right]^k
\end{aligned}$$

Raising both sides to the power $\frac{1}{k(k+1)}$, we get,

$$[(F_1 * F_2 * \cdots * F_k)]^{\frac{1}{k}} \geq \left[(F_1 * F_2 * \cdots * F_{k+1})\right]^{\frac{1}{k+1}} \tag{16}$$

which completes the induction step and hence the proof. □

We thus have the following theorem.

Theorem 4.4 *$\bar{H}$ as defined in (13) is IFRA if (12) holds.*

Proof The $\bar{P_k}$ defined in (14) satisfy the condition '$\bar{P_k}^{1/k}$ is decreasing in k' by virtue of the above lemma. The result now follows directly from Theorem 3.1. □

This seems to be an appropriate juncture to bring down the curtain on our brief overview of reliability shock models. We have tried, as far as practicable, to introduce the notion of shock models and to explain how they can give rise to various ageing classes (both monotonic and nonmonotonic). If this expository article has interested the reader sufficiently to pursue the subject of reliability shock models further, the authors would consider this venture a successful one.

Acknowledgements The authors are indebted to Professor Arnab K. Laha for his constant encouragement in making this article a reality and to Mr. Dhrubasish Bhattacharyya for some helpful discussions during the preparation of this manuscript.

References

A-Hameed, M., & Proschan, F. (1975). Shock models with underlying birth process. *Journal of Applied Probability, 12*(1), 18–28.

A-Hameed, M. S., & Proschan, F. (1973). Nonstationary shock models. *Stochastic Processes and their Applications, 1*, 383–404.

Abouammoh, A. M., & Ahmed, A. N. (1988). The new better than used failure rate class of life distributions. *Advances in Applied Probability, 20*, 237–240.

Abouammoh, A. M., Hendi, M. I., & Ahmed, A. N. (1988). Shock models with NBUFR and NBAFR survivals. *Trabajos De Estadistica, 3*(1), 97–113.

Akama, M., & Ishizuka, H. (1995). Reliability analysis of Shinkansen vehicle axle using probabilistic fracture mechanics. *JSME international journal. Ser. A, Mechanics and Material Engineering, 38*(3), 378–383.

Anderson, K. K. (1987). Limit theorems for general shock models with infinite mean intershock times. *Journal of Applied Probability, 24*(2), 449–456.

Anis, M. (2012). On some properties of the IDMRL class of life distributions. *Journal of Statistical Planning and Inference, 142*(11), 3047–3055.

Barlow, R. E., & Proschan, F. (1965). *Mathematical theory of reliability*. New York: Wiley.

Barlow, R. E., & Proschan, F. (1975). *Statistical theory of reliability and life testing*. Rinehart and Winston, New York: Holt.

Belzunce, F., Ortega, E.-M., & Ruiz, J. M. (2007). On non-monotonic ageing properties from the Laplace transform, with actuarial applications. *Insurance: Mathematics and Economics, 40*(1), 1–14.

Bogdanoff, J. L., & Kozin, F. (1985). *Probabilistic models of cumulative damage*. New York: Wiley.

Bryson, M. C., & Siddiqui, M. M. (1969). Some criteria for aging. *Journal of the American Statistical Association, 64*, 1472–1483.

Campean, I. F., Rosala, G. F., Grove, D. M., & Henshall, E. (2005). Life modelling of a plastic automotive component. In *Proceedings of Annual Reliability and Maintainability Symposium, 2005* (pages 319–325). IEEE.

Dasgupta, A., & Pecht, M. (1991). Material failure mechanisms and damage models. *IEEE Transactions on Reliability, 40*(5), 531–536.

Durham, S., & Padgett, W. (1997). Cumulative damage models for system failure with application to carbon fibers and composites. *Technometrics, 39*(1), 34–44.

Ebrahimi, N. (1999). Stochastic properties of a cumulative damage threshold crossing model. *Journal of Applied Probability, 36*(3), 720–732.

Esary, J. D., Marshall, A. W., & Proschan, F. (1973). Shock models and wear processes. *Annals of Probability, 1*, 627–649.

Feller, W. (1968). *An introduction to probability theory and its applications* (Vol. 1). New York: Wiley.

Garbatov, Y., & Soares, C. G. (2001). Cost and reliability based strategies for fatigue maintenance planning of floating structures. *Reliability Engineering & System Safety, 73*(3), 293–301.

Gertsbakh, I., & Kordonskiy, K. B. (2012). *Models of failure*. Springer Science & Business Media.

Glaser, R. E. (1980). Bathtub and related failure rate characterizations. *Journal of the American Statistical Association, 75*(371), 667–672.

Guess, F., Hollander, M., & Proschan, F. (1986). Testing exponentiality versus a trend change in mean residual life. *Annals of Statistics, 14*(4), 1388–1398.

Hollander, M., & Proschan, F. (1984). Nonparametric concepts and methods in reliability. In P. R. Krishnaiah & P. K. Sen (Eds.), *Handbook of statistics* (Vol. 4, pp. 613–655). Amsterdam: Elsevier Sciences.

Karlin, S. (1968). *Total positivity*, (Vol. 1). Stanford University Press.

Karlin, S., & Proschan, F. (1960). Pólya type distributions of convolutions. *The Annals of Mathematical Statistics*, 721–736.

Khan, R. A., Bhattacharyya, D., & Mitra, M. (2021). On classes of life distributions based on the mean time to failure function. *Journal of Applied Probability, 58*(2), in-press.
Klefsjö, B. (1981). HNBUE survival under some shock models. *Scandinavian Journal of Statistics, 8*, 39–47.
Klefsjö, B. (1983). A useful ageing property based on the Laplace transformation. *Journal of Applied Probability*, *20*, 615–626.
Kochar, S. C. (1990). On preservation of some partial orderings under shock models. *Advances in Applied Probability*, *22*(2), 508–509.
Lai, C. D., & Xie, M. (2006). *Stochastic ageing and dependence for reliability*. New York: Springer.
Loh, W. Y. (1984). A new generalization of the class of NBU distributions. *IEEE Transactions on Reliability, R-33*, 419–422.
Lukić, M., & Cremona, C. (2001). Probabilistic optimization of welded joints maintenance versus fatigue and fracture. *Reliability Engineering & System Safety*, *72*(3), 253–264.
Marshall, A. W., & Olkin, I. (2007). *Life distributions*. New York: Springer.
Mitra, M., & Basu, S. (1994). On a nonparametric family of life distributions and its dual. *Journal of Statistical Planning and Inference*, *39*, 385–397.
Mitra, M., & Basu, S. K. (1996). Shock models leading to non-monotonic ageing classes of life distributions. *Journal of Statistical Planning and Inference*, *55*, 131–138.
Nakagawa, T. (2007). *Shock and damage models in reliability theory*. Springer Science & Business Media.
Padgett, W. (1998). A multiplicative damage model for strength of fibrous composite materials. *IEEE Transactions on Reliability*, *47*(1), 46–52.
Pellerey, F. (1993). Partial orderings under cumulative damage shock models. *Advances in Applied Probability*, *25*(4), 939–946.
Pérez-Ocón, R., & Gámiz-Pérez, M. L. (1995). On the HNBUE property in a class of correlated cumulative shock models. *Advances in Applied Probability*, *27*(4), 1186–1188.
Petryna, Y. S., Pfanner, D., Stangenberg, F., & Krätzig, W. B. (2002). Reliability of reinforced concrete structures under fatigue. *Reliability Engineering & System Safety*, *77*(3), 253–261.
Rolski, T. (1975). Mean residual life. *Bulletin of the International Statistical Institute*, *46*, 266–270.
Satow, T., Teramoto, K., & Nakagawa, T. (2000). Optimal replacement policy for a cumulative damage model with time deterioration. *Mathematical and Computer Modelling*, *31*(10–12), 313–319.
Shanthikumar, J. G., & Sumita, U. (1983). General shock models associated with correlated renewal sequences. *Journal of Applied Probability*, *20*(3), 600–614.
Sobczyk, K., & Spencer, B, Jr. (2012). *Random fatigue: From data to theory*. Academic Press.
Sobczyk, K., & Trebicki, J. (1989). Modelling of random fatigue by cumulative jump processes. *Engineering Fracture Mechanics*, *34*(2), 477–493.
Yamada, K. (1989). Limit theorems for jump shock models. *Journal of Applied Probability*, *26*(4), 793–806.
Zacks, S. (1991). *Introduction to reliability analysis*. New York: Springer.

Explainable Artificial Intelligence Model: Analysis of Neural Network Parameters

Sandip Kumar Pal, Amol A. Bhave, and Kingshuk Banerjee

1 Introduction

In recent years, there has been growing interest of extracting patterns from data using artificial neural network (ANN)-based modelling techniques. The use of these models in the real-life scenarios is becoming primary focus area across different industries and data analytics practitioners. It is already established that the ANN-based models provide a flexible framework to build the models with increased predictive performance for the large and complex data. But unfortunately, due to high degree of complexity of ANN models, the interpretability of the results can be significantly reduced, and it has been named as "black box" in this community. For example, in banking system to detect the fraud or a robo-advisor for securities consulting or for opening a new account in compliance with the KYC method, there are no mechanisms in place which make the results understandable. The risk with this type of complex computing machines is that customers or bank employees are left with a series of questions after a consultancy or decision which the banks themselves cannot answer: "Why did you recommend this share?", "Why was this person rejected as a customer?", "How does the machine classify this transaction as terror financing or money laundering?". Naturally, industries are more and more focusing on the transparency and understanding of AI when deploying artificial intelligence and complex learning systems.

S. K. Pal (✉) · A. A. Bhave
Cognitive Business and Decision Support, IBM India, Bengaluru, India
e-mail: sandipkumar.pal@gmail.com

A. A. Bhave
e-mail: amobhave@in.ibm.com

K. Banerjee
Cognitive Computing and Analytics, IBM Global Business Services, Bengaluru, India
e-mail: kingshukb@in.ibm.com

A. K. Laha (ed.), *Applied Advanced Analytics*, Springer Proceedings in Business and Economics, https://doi.org/10.1007/978-981-33-6656-5_4

Probably, this has opened a new direction of research works to develop various approaches to understand the model behaviour and the explainability of the model structure. Recently, Joel et al. (2018) has developed explainable neural network model based on additive index models to learn interpretable network connectivity. But it is not still enough to understand the significance of the features used in the model and the model is well specified or not.

In this article, we will express the neural network (NN) model as nonlinear regression model and use statistical measures to interpret the model parameters and the model specification based on certain assumptions. We will consider only multilayer perceptron (MLP) networks which is a very flexible class of statistical procedures. We have arranged this article as: (a) explain the structure of MLP as feed-forward neural network in terms of nonlinear regression model, (b) the estimation of the parameters, (c) properties of parameters and their asymptotic distribution, (d) simulation study and conclusion.

2 Transparent Neural Network Model (TRANN)

In this article, we have considered the MLP structure given in Fig. 1. Each neural network can be expressed as a function of explaining variable $X = \left[x_1, x_2, \ldots, x_p\right]$ and the network weights $\omega = \left(\gamma', \beta', b'\right)$ where α' is the weights between input and hidden layers, β' is the weights between hidden and output layers and b' is the bias of the network. This network is having the following functional form

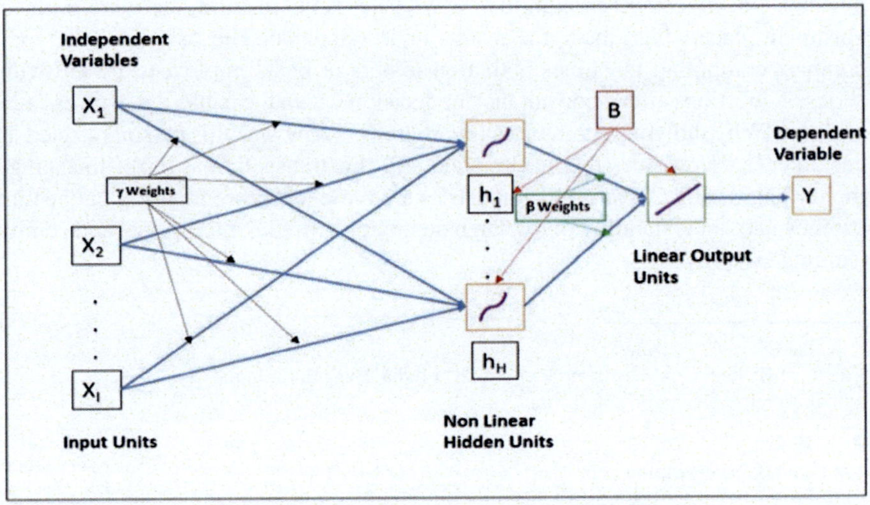

Fig. 1 A multilayer perceptron neural network: MLP network with three layers

$$F(X, \omega) = \sum_{h=1}^{H} \beta_h g(\sum_{i=1}^{I} \gamma_{hi} x_i + b_h) + b_{00} \tag{1}$$

where the scalars I and H denote the number of input and hidden layers of the network and g is a nonlinear transfer function. The transfer function g can be considered as either logistic function or the hyperbolic tangent function. In this paper, we have considered logistic transfer function for all the calculation. Let us assume that Y is dependent variable and we can write Y as a nonlinear regression form

$$Y = F(X, \omega) + \epsilon \tag{2}$$

where ϵ is $i.i.d$ normal distribution with $E[\epsilon] = 0$, $\mathrm{E}[\epsilon\epsilon'] = \sigma I$. Now, Eq. (2) can be interpreted as parametric nonlinear regression of Y on X. So based on the given data, we will be able to estimate all the network parameters. Now the most important question is what would be the right architecture of the network, how we can identify the number of hidden units in the network and how to measure the importance of those parameters. The aim is always to identify an optimum network with small number of hidden units which can well approximate the unknown function (Sarle 1995). Therefore, it is important to derive a methodology not only to select an appropriate network but also to explain the network well for a given problem.

In the network literature, available and pursued approaches are *regularization*, *stopped-training* and *pruning* (Reed 1993). In *regularization* methods, we can minimize the network error (e.g. sum of error square) along with a penalty term to choose the network weights. In the *stopped-training* data set, the training data set split into training and validation data set. The training algorithm is stopped when the model errors in the validation set begin to grow during the training of the network, basically stopping the estimation when the model is overparameterized or overfitted. It may not be seen as sensible estimates of the parameters as the growing validation error would be an indication to reduce the network complexity. In the *pruning* method, the network parameters are chosen based on the "significant" contribution to the overall network performance. However, the "significance" is not judged by based on any theoretical construct but more like a measure of a factor of importance.

The main issue with *regularization*, *stopped-training* and *pruning* is that they are highly judgemental in nature which makes the model building process difficult to reconstruct. In transparent neural network (TRANN), we are going to explain the statistical construct of the parameters' estimation and their properties through which we explain the statistical importance of the network weights and will address well the model misspecification problem. In the next section, we will describe the statistical concept to estimate the network parameters and their properties. We have done a simulation study to justify our claim.

3 TraNN Parameter Estimation

In general, the estimation of parameters of a nonlinear regression model cannot be determined analytically and needs to apply the numerical procedures to find the optima of the nonlinear functions. This is a standard problem in numerical mathematics. In order to estimate the parameters, we minimized squared error, $SE = \sum_{t=1}^{T}(Y_t - F(X_t, \omega))^2$, and applied backpropagation method to estimate the parameters. Backpropagation is the most widely used algorithm for supervised learning with multi-layered feed-forward networks. The repeated application of chain rule has been used to compute the influence of each weight in the network with respect to an error function SE in the backpropagation algorithm (Rumelhart et al. 1986) as:

$$\frac{\partial SE}{\partial \omega_{ij}} = \frac{\partial SE}{\partial s_i}\frac{\partial s_i}{\partial \text{net}_i}\frac{\partial \text{net}_i}{\partial \omega_{ij}} \tag{3}$$

where ω_{ij} is the weight from neuron j to neuron i, s_i is the output, and net_i is the weighted sum of the inputs of neuron i. Once the partial derivatives of each weight are known, then minimizing the error function can be achieved by performing

$$\breve{\omega}_{t+1} = \breve{\omega}_t - \eta_t[-\nabla F(X_t, \breve{\omega}_t)]'[Y_t - F(X, \breve{\omega}_t)], t = 1, 2, \ldots, T \tag{4}$$

Based on the assumptions of the nonlinear regression model (2) and under some regularity conditions for F, it can be proven (White 1989) that the parameter estimator $\hat{\omega}$ is consistent with asymptotic normal distribution. White ((White, 1989)) had shown that the parameter estimator an asymptotically equivalent estimator can be obtained from the backpropagation estimator using Eq. (4) when η_t is proportional to t^{-1} as

$$\hat{\omega}_{t+1} = \breve{\omega}_t + \left[\sum_{t=1}^{T}\nabla F(X_t, \breve{\omega}_t)'\nabla F(X_t, \breve{\omega}_t)\right]^{-1} \tag{5}$$
$$\times \sum_{t=1}^{T}\nabla F(X_t, \breve{\omega}_t)'[Y_t - F(X, \breve{\omega}_t)], t = 1, 2, \ldots, T$$

In that case, the usual hypothesis test like Wald test or the LM test for nonlinear models can be applied. Neural network belongs to the class of misspecified models as it does not map to the unknown function exactly but approximates. The application of asymptotic standard test is still valid as the misspecification can be taken care through covariance matrix calculation of the parameters (White 1994). The estimated parameters $\hat{\omega}$ are normally distributed with mean ω^* and covariance matrix $\frac{1}{T}C$. The parameter vector ω^* can be considered as best projection of the misspecified model onto the true model which lead to:

$$\sqrt{T}(\hat{\omega} - \omega^*) \sim N(0, C) \tag{6}$$

where the T denotes the number of observations. As per the theory of misspecified model (Anders 2002), the covariance matrix can be calculated as

$$\frac{1}{T} = A^{-1}BA^{-1} \tag{7}$$

where the matrix A and B can be expressed as $A \equiv E[\nabla^2 SE_t]$ and $B \equiv E[\nabla SE_t \nabla SE_t^{'}]$. SE_t denotes the squared error contribution of the tth observations, and ∇ is the gradient with respect to the weights.

4 TRANN Model Parameter Test for Significance

The hypothesis tests for significance of the parameters are an instrument for any statistical models. In TRANN, we are finding and eliminating redundant inputs from the feed-forward single layered network through statistical test of significance. This will help to understand the network well and will be able to explain to network connection with mathematical evidence. This will help to provide a transparency to the model as well. The case of irrelevant hidden units occurs when identical optimal network performance can be achieved with fewer hidden units. For any regression method, the value of t-statistic plays an important role for hypothesis testing whereas it is overlooked in neural networks. The non-significant parameters can be removed from the network, and the network can be uniquely defined (White 1989). This is valid for linear regression as well as neural networks. Here, we estimate the t-statistic as

$$\frac{\hat{\omega}_k - \omega_{H_0}(k)}{\hat{\sigma}_k} \tag{8}$$

where $\omega_{H_0}(k)$ denotes the value or the restrictions to be tested under null hypothesis H_0. The $\hat{\sigma}_k$ is the estimated standard deviation of the estimated parameter $\hat{\omega}_k$. Later, we have estimated the variance–covariance matrix $\hat{C}$ where the diagonal elements are ω_k and the $\hat{C}$ can be estimated as

$$\frac{1}{T}\hat{C} = \hat{A}^{-1}\hat{B}\hat{A}^{-1} \tag{9}$$

$$\hat{A}^{-1} = \frac{1}{T}\sum_{t=1}^{T}\frac{\partial^2 SE_t}{\partial\hat{\omega}\partial\hat{\omega}'} \text{ and } \hat{B}^{-1} = \sum_{t=1}^{T}\hat{\epsilon}_t^2(\frac{\partial F(t,\hat{\omega})}{\partial\hat{\omega}})(\frac{\partial F(t,\hat{\omega})}{\partial\hat{\omega}})' \tag{10}$$

where $\hat{\epsilon}_t^2$ is the square of estimated error for tth sample.

Equation (6) implies that asymptotic distribution of the network parameters is normally distributed and it is possible to perform the test of significance of each parameter using the estimated covariance matrix $\hat{C}$. Then, both Wald test and LM test are applicable as per the theory of misspecified model (Anders 2002).

5 Simulation Study

We have performed a simulation study to establish the estimation methods and hypothesis test of significance with a 8-2-1 feed-forward network where we have considered eight input variables, one hidden layer with two hidden units and one output layer. Therefore, as per the structure of Eq. (1), the network model contains 21 parameters and we have set the parameter values as $b' = (b_{00} : 0.91, b_1 : -0.276, b_2 : 0.276)$
$\beta' = (\beta_1 : 0.942, \beta_2 : 0.284)$
$\gamma' = (\gamma_{11} = -1.8567, \gamma_{21} = -0.0185, \gamma_{31} = -0.135), \gamma_{41} = 0.743, \gamma_{51} = 0.954, \gamma_{61} = 1.38, \gamma_{71} = 1.67, \gamma_{81} = 0.512, \gamma_{12} = 1.8567, \gamma_{22} = 0.0185, \gamma_{32} = 0.135, \gamma_{42} = -0.743, \gamma_{52} = -0.954, \gamma_{62} = -1.38, \gamma_{72} = -1.67, \gamma_{82} = -0.512)$
and the error term ϵ is generated from normal distribution with mean zero and standard deviation 0.001. In the model, the independent variables $X = [x_1, ..., x_8]$ are drawn from exponential distribution. We have generated 100,000 samples using the above parameters, and then we have taken multiple sets of 5000 random sample of observations out of 100,000 observations and derived the estimates of the parameters and confidence intervals. We are calling this method as bootstrap method. The estimated values of the parameters, standard errors, confidence interval, t-values and p-values through bootstrapping method are given in Table 1. The results based on the asymptotic properties of the estimates are given in Table 2 based on Eq. (9). Both the methods are establishing the test of significance of parameters under null hypothesis $H_0 : \omega = 0$.

Table 1 Results using bootstrapping method

Coefficients	Estimates	Std. error	t value	95% C.I.	Pr[> |t|]
b_{00}	0.917	1.089E−03	842.057	[0.916 , 0.920]	< 0.001
b_1	−0.283	4.613E−03	−61.348	[−0.289 , −0.272]	< 0.001
b_2	0.300	1.312E−02	22.866	[0.284 , 0.327]	< 0.001
β_1	0.936	1.433E−03	653.175	[0.934 , 0.938]	< 0.001
β_2	0.279	1.086E−03	256.906	[0.278 , 0.281]	< 0.001
γ_{11}	−1.854	7.246E−03	−255.865	[−1.861, −1.833]	< 0.001
γ_{21}	−0.025	3.732E−03	−6.699	[−0.031, −0.017]	< 0.001
γ_{31}	−0.142	3.050E−03	−46.557	[−0.147 , −0.137]	< 0.001
γ_{41}	0.736	3.217E−03	228.785	[0.731 , 0.741]	< 0.001
γ_{51}	0.947	4.142E−03	228.634	[0.941 , 0.952]	< 0.001
γ_{61}	1.373	4.771E−03	287.78	[1.365 , 1.380]	< 0.001
γ_{71}	2.133	3.488E−02	61.153	[2.075 , 2.182]	< 0.001
γ_{81}	0.019	3.294E−03	5.768	[0.015 , 0.025]	< 0.001
γ_{12}	1.873	2.402E−02	77.977	[1.816 , 1.910]	< 0.001
γ_{22}	0.042	1.401E−02	2.998	[0.025 , 0.070]	< 0.001
γ_{32}	0.160	1.233E−02	12.976	[0.143 , 0.185]	< 0.001
γ_{42}	−0.730	1.081E−02	−67.53	[−0.746 , -0.705]	< 0.001
γ_{52}	−0.941	1.347E−02	−69.859	[−0.959 , −0.911]	< 0.001
γ_{62}	−1.375	1.544E−02	−89.054	[−1.398 , −1.341]	< 0.001
γ_{72}	−2.039	1.333E−01	−15.296	[−2.197 , −1.796]	< 0.001
γ_{82}	−0.065	1.170E−02	−5.556	[−0.08 , −0.042]	< 0.001

Table 2 Results using asymptotic properties

Coefficients	Estimates	Std. error	t value	Pr[> |t|]
b_{00}	0.917	1.646E−04	5573.017	<0.001
b_1	−0.283	3.715E−04	−761.616	<0.001
b_2	0.305	1.356E−03	224.886	<0.001
β_1	0.936	1.657E−04	5649.898	<0.001
β_2	0.279	1.646E−04	1695.213	<0.001
γ_{11}	−1.855	7.034E−05	−26371.921	<0.001
γ_{21}	−0.025	4.436E−05	−574.123	<0.001
γ_{31}	−0.142	3.441E−05	−4125.615	<0.001
γ_{41}	0.732	2.823E−05	25,930.712	<0.001
γ_{51}	0.943	4.694E−05	20,096.352	<0.001
γ_{61}	1.373	7.827E−05	17,541.977	<0.001
γ_{71}	2.151	9.378E−05	22,930.994	<0.001
γ_{81}	0.021	1.652E−05	1300.632	<0.001
γ_{12}	1.885	2.446E−04	7704.932	<0.001
γ_{22}	0.042	1.637E−04	255.681	<0.001
γ_{32}	0.161	1.273E−04	1266.425	<0.001
γ_{42}	−0.717	1.077E−04	−6654.568	<0.001
γ_{52}	−0.932	1.749E−04	−5325.529	<0.001
γ_{62}	−1.377	2.837E−04	−4855.127	<0.001
γ_{72}	−2.103	3.496E−04	−6016.115	<0.001
γ_{82}	−0.075	5.922E−05	−1265.38	<0.001

6 Conclusion

Neural networks are a very flexible class of assumptions about the structural form of the unknown function F. In this paper, we have used nonlinear regression technique to explain the network through statistical analysis. The statistical procedures usable for model building in neural networks are significance test of parameters through which an optimal network architecture can be established. In our opinion, the transparent neural network is a major requirement to perform a diagnosis of neural network architecture which not only approximates the unknown function but also explains the network features well through the statistical nonlinear modelling assumptions. As a next step, we would like to investigate more on the deep neural networks based on the similar concepts.

Acknowledgements We use this opportunity to express our gratitude to everyone who supported us in this work. We are thankful for their intellectual guidance, invaluable constructive criticism and friendly advice during this project work. We are sincerely grateful to them for sharing their truthful and illuminating views on a number of issues related to the project. We express our warm thanks to our colleagues Koushik Khan and Sachin Verma for their support to write code in Python and R. We would also like to thank Prof. Debasis Kundu from IIT Kanpur who provided the valuable references and suggestions for this work.

References

Anders, U. (2002). Statistical model building for neural networks. In *963 Statistical Model Building for Neural Networks*.

Joel, V., et al. (2018). Explainable neural networks based on additive index model. arXiv.

Reed, R. (1993). Pruning algorithms—A survey. *IEEE Transactions on Neural Networks*, *4*, 740–747.

Rumelhart, D. E., et al. (1986). A direct adaptive method for faster backpropagation learning-the rprop algorithm. Parallel distributed Processing.

Sarle, W. S. (1995). Stopped training and other remedies for overfitting. In *Proceedings of the 27th Symposium on the Interface*.

White, H. (1994). *Estimation, inference and specification analysis*. Cambridge University Press.

White, H. (1989). Learning in neural networks: A statistical perspective. *Neural Computation*, *1*, 425–464.

Style Scanner—Personalized Visual Search and Recommendations

Abhishek Kushwaha, Saurav Chakravorty, and Paulami Das

1 Introduction

Fashion is driven more by visual appeal than by any other factor. We all buy clothes that we think 'looks good' on us. Today, a large portion of e-commerce is driven by fashion and home décor which includes apparels, bags, footwear, curtains, sofas, etc. Any search engine that does not have the capability to 'see' will not work in a way that humans do. The need for visual search becomes the next phase of evolution of e-commerce search engines.

Present e-commerce search engines lack in this regard. They use metadata to look for similarity between clothes which does not capture finer details of the apparel and design. We see several recommendation engines that use collaborative filtering to recommend apparels and other products. But collaborative filtering cannot understand the nuances of visual design as perceived by humans.

To understand the importance of visual search, let us understand the typical stages of buying fashion products. Whether one buys offline or online, human behaviour tends to remain the same. On a digital platform, it gets restricted by the features of the platform. When one decides to buy clothes, they have a category such as shirt, tee or dress in mind. But details of design and features are rarely pre-decided.

For shopping offline or online, the first stage is exploration. Several designs and patterns are explored before narrowing down the search to few designs. The second

A. Kushwaha (✉) · S. Chakravorty · P. Das
Brillio Technologies, Bengaluru, India
e-mail: abhishek.k@brillio.com
URL: http://www.brillio.com

S. Chakravorty
e-mail: saurav.chakravorty@brillio.com

P. Das
e-mail: paulami.das@brillio.com

A. K. Laha (ed.), *Applied Advanced Analytics*, Springer Proceedings in Business and Economics, https://doi.org/10.1007/978-981-33-6656-5_5

stage begins at this point when one investigates their chosen designs in greater detail. At this second stage, given the exploratory nature of people, there is a tendency to look for clothes that are similar to the chosen one. In offline stores, this is evinced from customers asking the sales person for variations of the chosen design—clothes with a different colour, a slight variation in design, same pattern with different borders, etc. The transition from first to second stage is bidirectional, meaning, after looking for similar clothes the person may again go back to exploration stage. And the process continues.

On e-commerce platforms, for stage one, retrieval algorithm should focus on keeping high variance. This means the first search should show a variety of clothes, not similar ones. One can also add taste, body size, etc., to this algorithm. For second stage, the search algorithm should retrieve choices that look similar to the ones chosen in the first stage. This helps in faster decision-making. Through this, the customer already has a selection of similar items to choose from and therefore can focus on the final selection.

2 Related Work

Many models have been developed in the past to capture the notion of similarity. Image features are extracted using traditional computer vision algorithms like SIFT Lowe (1999) and HOG Dalal and Triggs (2005), and on top of them image similarity models are learned. These have been studied in Boureau et al. (2010), Chechik et al. (2010), Taylor et al. (2011). Traditional computer vision (CV) techniques have limited expressive power and hence have not worked very well. With recent developments in neural network, deep CNNs have been used with great success for object recognition Krizhevsky et al. (2012), Simonyan et al. (2014), Szegedy et al. (2015), detection, etc. They have been able to take CV to the next level where they have become effective enough to be applied to real-life problems with great results. In CNNs, successive layers learn to represent the image with increasing level of abstraction. The final layer contains abstract descriptor vector. CNN is robust to illumination variation, in-image location variation, occlusion, etc. Deep CNNs have been made successful with residual networks He et al. (2016) and have been able to truly learn image features.

In real-life applications, we may not be interested in the similarity between cats and dogs but we are interested in similarity between two dogs. Similarly, for fashion search, we are interested in similarities in various aspects of products such as print designs. To be able to recognize an object, the final layer must understand the image at the highest level of abstraction (cat, dog, horse). But the notion of similarity lies at the lower levels.

Therefore, similarity models require lower-level CNN feature abstraction as output too . Learning finer details along with abstract feature has been studied by Wang et al. (2014). We will make use of this architecture for improved learning.

Siamese network has also been used with contrastive loss by Chopra et al. (2005), for similarity assessment. In this architecture, there are two CNNs with shared weights with binary output. This helps to find similarity between two clothes but does not capture fine-grained similarity. These networks have proved to be good for certain face verification tasks but not for design similarity.

Wang et al. (2016) propose a deep Siamese network with a modified contrastive loss and a multitask network fine-tuning scheme. Their model being a Siamese network also suffers from the same limitations discussed above.

Image similarity using triplet networks has been studied in Wang et al. (2014), Lai et al. (2015). The pairwise ranking paradigm used here is essential for learning fine-grained similarity. Further, the ground truth is more reliable as it is easier to label relative similarity than absolute similarity.

3 Our Approach

We built this tool as an enterprise solution. We soon realized the same solution would not work with all enterprises. Each enterprise will have different data. For example, one may have metatag level info for each apparel but others might not. It is a difficult task to generate such a huge amount of data manually. Hence, building different models became necessary.

Our core approach has been to learn embeddings which captures the notion of similarity. We achieved this using CNNs that learn visual feature with various types of output architecture and loss function depending upon the type/level of data available. We built three models. The first one was with least amount of data. We had category-level information about the clothes (shirt, shorts, tees, tops, etc.). Using this, we designed a classification model with Softmax loss and used it to get embeddings from the second last layer after training. The idea was that the Softmax loss would cluster the same category clothes and within each cluster the intra-class distribution would be in accordance with similarity.

The second model we built was using metadata tags for each item. We modelled this as tag detection problem (multi-class and multi-label) with sigmoid loss and used the embeddings from second last layer to find similarity. This model performed much better than the previous one, not because of the loss function, but because it had more data for supervised learning.

The third one we built was modelled on Wang et al. (2014). In this deep CNN architecture, two more shallow branches were added to capture fine-grained similarity. The deep branch we chose was Inception and VGG16, Fig. 1. We found that VGG16 performed better than Inception. The training is based on triplet approach with ranking loss. This significantly improved the model performance as ranking loss objective is the same as ranking clothes based on similarity. The dataset creation for triplet loss is complicated because of the subjective nature of similarity. However, it is very simple in case of learning embeddings for face recognition system (as the two photographs are either of same person or of different persons, i.e. no subjectivity).

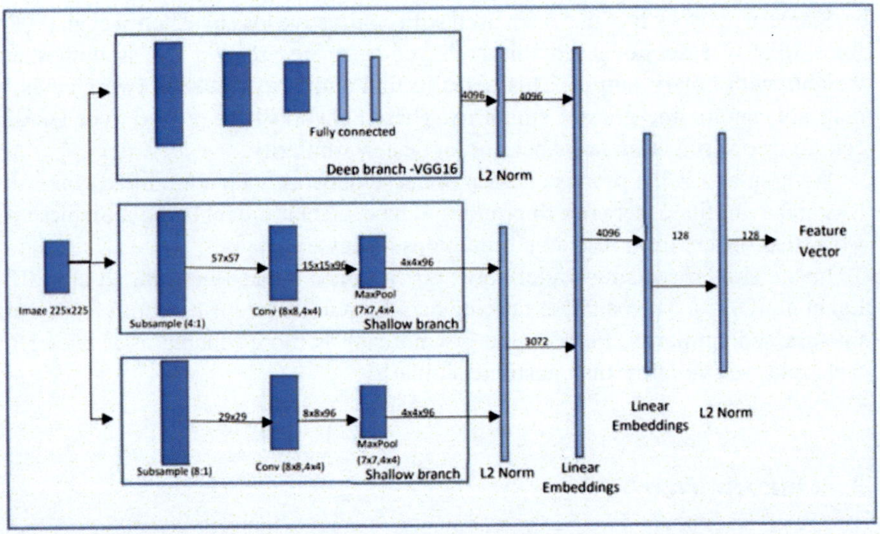

Fig. 1 Multi-scale network architecture. Each image goes through one deep and two shallow branches. The number shown on the top of an arrow is the size of the output image or feature. The number shown on the bottom of a box is the size of the kernels for the corresponding layer

For the triplet loss dataset, each training data element consists of 3 images <a,p,n>, an anchor image <a>, a positive image <p>, a negative image <n>. <a,p> pair of images are expected to be more visually similar than <a,n>. So, the dataset needs to be labelled for relative similarity than absolute similarity. Although this is not simple, it is doable.

We use two types of triplets, in-class and out-of-class triplets. The out-of-class triplet has easy negatives which helps to learn coarse-grained differences. This dataset is easy to make. The in-class triplet has hard negatives, meaning visually on broad levels they look like the anchor image but on finer level are not that similar (different stripe widths, etc.). This helps in learning fine-grained distinctions, and the model becomes sensitive to difference in pattern and colour. In-class triplet is harder to create, and its quality affects the performance of the final model.

4 Training Dataset Generation

For our first model, the classification model, dataset is comprised of catalogue images taken in professional studio lighting conditions. The labels corresponded to their category. For the second model, the multi-class, multi-label model, dataset consisted of catalogue images with metatags (like stripe, blue, short, etc.). Each image had several tags defining its visual characteristics. This has been manually labelled. We chose this model as we had data available. But if this metadata is not available, this model cannot be used.

For the triplet model, as described earlier, data creation is not straightforward, and one cannot do it manually for millions of image (metadata creation has been the work of many years). We adopted an algorithmic way of labelling data for relative similarity. For this, we create multiple basic similarity models (BMSs) that focus on various aspects of similarity. Few of them are as follows: Our first model served as a BMS that focusses on courser details. Others were ColorHist which is LAB colour histogram of image foreground and PatternNet which is model trained to recognize pattern (stripes, checks, etc.) for which we used metadata. We also used our second model as one of the BMSs. These BMSs were built using metadata information, and the type of BMS used was dependent on the information available.

To form a triplet, an anchor image a is selected, and a positive image is randomly selected from a set of 200 positive images. This set is formed by the following process: each BMS identifies 500 nearest neighbours to anchor image a, and top 200 from the union of all BMS is taken as the set of positive images. For in-class negative image, union of all BMS image ranked between 500 and 1000 is taken as sample set and from this an image is randomly selected for in-class negative. Out-class negative sample space is rest of the universe within the category group. The final set contained 20% in-class negative and 80% out-class negatives. For this exercise, these numbers were arbitrarily chosen, but further fine-tuning can be done to find the optimal ratio.

5 Implementation Details

Classification model had deep–shallow network. Multi-label, multi-class model had deep Inception model Szegedy et al. (2015). For the triplet model, we used deep–shallow network. All the models were trained on 0.2 million images. The final dataset for triplet-based training was around 1 million.

The CNN architecture was implemented in Keras and TensorFlow. Training was done in Nvidia GPUs on Microsoft Azure platform.

6 Dataset and Evaluation

For development, we used a dataset provided by clients. It consisted of 2 lakh images from the apparel category. This included shirts, T-shirts, tops, dresses. Images had metadata too.

We created triplets programmatically. Thousand of them were manually checked, and incorrect ones were removed. Then, the models were evaluated on this dataset basis, the percentage of triplets that were correctly ranked by them (i.e. given a triplet <a,p,n>, if D(a,p)<D(a,n), where $D(x, y)$ is defined as Euclidean distance between vector x and y, then we count it as success otherwise fail). The results are shown in Table 1.

Table 1 In-class and out-of-class triplet accuracy

Method	In-class triplet accuracy	Out-of-class triplet accuracy
Classification model	65.71	74.38
Multi-class, multi-label	77.63	89.54
PatternNet	60.15	83.22
Ranking network	91.52	97.79

7 Production Pipeline

We productionized the whole system to cater to end user needs. There were trade-offs involved between cost and quality. Our present components include:

Embedding calculation service: We created CPU-based service with elastic load balancer to calculate the embedding vector for new items added from second last layer of our CNN. The GPU is not used for this as it is not cost effective. GPU is used only for training. These embeddings were saved in a distributed file system (Azure).

Nearest neighbour search: We used open-source annoy library for finding nearest neighbours. We did not apply approximate nearest neighbour technique, locality sensitive hashing (LSH), as we found it significantly decreases model performance. We kept our embedding vector dimension at 128. Although higher-dimensional embeddings were giving better results, they drastically increased execution time of nearest neighbour search.

We calculated nearest neighbour for new items added and updated that for the past ones. We also needed to update it whenever an item was deleted.

Reduction in search space: Since we had metadata (shirt, shorts), (male, female), etc., for each image, we effectively used this to reduce our search space for faster execution.

8 Future Developments

In one of our latest experiments, we have observed that training the model with removed background enhances model performance. This is one of the areas we are working on for further improvement. The background removal must be automatic and accurate. GrabCut Rother et al. (2004), an image segmentation method, can be used for background removal semiautomatically. By using data created by this technique, we can train one more CNN model to automatically remove the background.

9 Conclusion

We presented an application of computer vision technology for e-commerce space that supports the intuitive shopping behaviour and thereby improves user experience and business performance. We have explained the architecture of various types of models and the corresponding data generation process. We have also noted the trade-offs we made while designing the system. There are several improvements that can be done, and we at Brillio are constantly developing new techniques to improve the system quality and execution time.

References

Boureau, Y.-L., Bach, F., LeCun, Y., & Ponce, J. (2010). Learning mid-level features for recognition. In *2010 IEEE Computer Society Conference on Computer Vision and Pattern Recognition* (pp. 2559–2566). IEEE.

Chechik, G., Varun S., Uri, S., & Samy, Bengio. (2010). Large scale online learning of image similarity through ranking. *Journal of Machine Learning Research, 11*(3).

Chopra, S., Raia, H., & Yann, L. (2005). Learning a similarity metric discriminatively, with application to face verification. In *2005 IEEE Computer Society Conference on Computer Vision and Pattern Recognition (CVPR'05)* (Vol. 1, pp. 539–546). IEEE.

Dalal, N., & Triggs, B. (2005). Histograms of oriented gradients for human detection. In *Computer Vision and Pattern Recognition. CVPR 2005* (pp. 886–893). IEEE Computer Society Conference on 1.

He, K., Xiangyu, Z., Shaoqing, R., & Jian, S. (2016). Deep residual learning for image recognition. In *Proceedings of the IEEE Conference on Computer Vision and Pattern Recognition* (pp. 770–778).

Krizhevsky, A., Ilya, S., & Hinton, G. E. (2012). Imagenet classification with deep convolutional neural networks. In *Advances in neural information processing systems* (pp. 1097–1105).

Lai, H., Yan, P., Ye, L., & Shuicheng, Y. (2015). Simultaneous feature learning and hash coding with deep neural networks. In *Proceedings of the IEEE Conference on Computer Vision and Pattern Recognition* (pp. 3270–3278).

Lowe, D. G. (1999). *Object recognition from local scale-invariant features*. Paper presented at the meeting of the Proceedings of the International Conference on Computer Vision ICCV, Corfu.

Rother, Carsten, Kolmogorov, Vladimir, & Blake, Andrew. (2004). " GrabCut" interactive foreground extraction using iterated graph cuts. *ACM Transactions on Graphics (TOG)*, *23*(3), 309–314.

Simonyan, K., & Andrew, Z. (2014). Very deep convolutional networks for large-scale image recognition. arXiv preprint arXiv:1409.1556.

Szegedy, C., Wei, L., Yangqing, J., Pierre, S., Scott, R., Anguelov, D., Erhan, D., et al. (2015). Going deeper with convolutions, 1–9. In *IEEE Conference on Computer Vision and Pattern Recognition (CVPR)*. IEEE, New Jersey.

Taylor, G. W., Spiro, I., Christoph, B., & Rob, F. (2011). Learning invariance through imitation. In *CVPR 2011* (pp. 2729–2736). IEEE.

Wang, X., Sun, Z., Zhang, W., Zhou, Y., & Jiang, Y.-G. (2016). Matching user photos to online products with robust deep features. In *Proceedings of the 2016 ACM on International Conference on Multimedia Retrieval* (pp. 7–14).

Wang, J., Yang, S., Thomas, L., Chuck, R., Jingbin, W., James, P., et al. (2014). Learning fine-grained image similarity with deep ranking. In *Proceedings of the IEEE Conference on Computer Vision and Pattern Recognition* (pp. 1386–1393).

Artificial Intelligence-Based Cost Reduction for Customer Retention Management in the Indian Life Insurance Industry

Sanjay Thawakar and Vibhu Srivastava

1 Introduction

1.1 Company Information

Max Life Insurance (Max Life) is the largest non-bank private life insurer in India with total revenue of Rs. 12,501 crore and assets of Rs. 52,237 crore in FY 2017–18. Max Life Insurance is a joint venture between Max India Ltd. and Mitsui Sumitomo Insurance Co., Ltd. It offers comprehensive life insurance and retirement solutions for long-term savings and protection to 3.5 million customers. It has a countrywide diversified distribution model including the agent advisors, exclusive arrangement with Axis Bank and several other partners.

1.2 Background

Unlike many other industries, customer retention pays a very critical for life insurance companies. On an average, life insurers become profitable only after 6–7 renewal premiums are paid by the customer. Customer retention efficiency is defined by persistency ratio, which is one of the most important metrics for the life insurance companies. Persistency is measured by the percentage of policies renewed year on year.

According to the Insurance Regulatory and Development Authority of India (IRDAI), insure@nceiknowledge admin experts speak (2018), in 2015–16, the average persistency rate for life insurance policies in 13th month after issuance was 61%, indicating that only 61% policies paid their first renewal premium as on

S. Thawakar · V. Srivastava (✉)
Analytics Centre of Excellence, Max Life Insurance, New Delhi, India
e-mail: vibhu182000@yahoo.com

A. K. Laha (ed.), *Applied Advanced Analytics*, Springer Proceedings in Business and Economics, https://doi.org/10.1007/978-981-33-6656-5_6

13th month. Globally, the persistency is around 90% in the 13th month and over 65% after 5 years of policy issuance, while the acceptable percentage of persistency in life insurance is 80% for polices after 3 years and 60% after 10 years.

With the stringent persistency guidelines issued by IRDAI to protect policy holder's interests, retention management has become a key focus area for the life insurance companies.

According to a survey conducted by data aggregation firm LexisNexis Risk Solutions (Jones 2017), factors including lack of need-based selling, lack of identification of changing needs of the customer, limited payment reminders and inefficiencies in retention management contribute to lower than benchmark persistency for life insurers in India. Life insurance companies are spending top dollars on customer retention management, and improving persistency along with reduced retention costs has become a key focus area.

1.3 Purpose of Research

Max Life has a base of nearly *3.5 million* active policy holders, comprising freshly acquired policy holders and existing policy holders. Income from renewal premiums accounts for 70% of the total revenue, representing the future cash flows, and plays an important role in company's profitability and valuation.

For Max Life, the renewal income collection and persistency are managed by the company's *customer retention* operations. Max Life's customer retention operations primarily include reaching out to customers through telephonic renewal calls, or other mediums such as mobile text messages, to remind them to pay renewal premiums on time. This retention effort is made to maximize both persistency and renewal premium collection. Figure 1 shows the overview of Max Life business operations along with key activities of customer retention operations.

As the policy tenure increases, the retention rates keep decreasing and thus significant effort is required to retain customers over time. There are many different reasons contributing to the discontinuation of policies. Within Max Life, these factors are

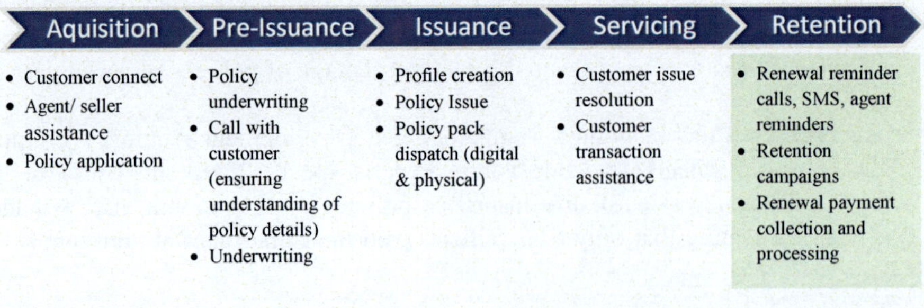

Fig. 1 Overview of Max Life business operations and retention activities

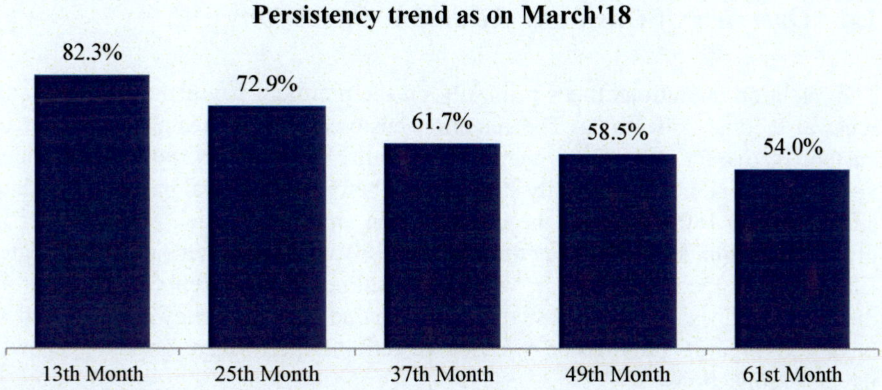

Fig. 2 Max Life persistency based on policy vintage

categorized differently based on type of policy discontinuation. For example, a policy is considered to be ***lapsed*** if the customer did not pay renewal premium for a period of 180 days from due date, but if opt to ***surrender*** the policy, the company has to pay some surrender amount to the customer. In both these cases, policy will be discontinued and will be termed as nonpersistent; however, the reason for discontinuation is different.

For the purpose of the research, only lapsed polices are considered as lapse events contribute to 85% of the nonpersistent policies and largely, the proactive retention efforts are possible only for lapse events. Within the *lapsed* policies, a further subclassification defines different types of lapses, and this subclassification is defined separately in the model building section.

Typically in the entire renewal book (all policies due for renewals in a year), *88% of the policies are retained and 12%* of the policies are *lapsed* every year. Every 1% policy discontinuation rate represents Rs 100 Cr of renewal income. Figure 2 shows the current persistency rates for MLI policy base.

Retention costs, driven by renewal calls, are a major cost head for the company. As the renewal book is continuously increasing, driven by a sharp 15–20% sales growth, the retention costs are proportionally increasing, reducing company's profitability.

The purpose of the research was to use data and design an advanced analytics solution to optimize customer retention costs and help reduce lapse rates gradually. The objective of the analytical solution was to classify the customers based on propensity of lapsation and using this propensity score to

- Optimize the customer retention cost through reduction in renewal calls, without impacting the renewal income.
- Create a segmented renewal strategy to match the customer segment with appropriate renewal efforts and leverage other communication modes like SMS/digital, thereby improving retention rates.

1.4 Overview of Customer Retention Operations

The retention operations focus primarily on the premium payment reminders and renewal income collections. The retention customer contact team reaches out to customers, due for renewal payment through calls, SMS, emails and physical channels. The reminder calls typically start 15 days before the renewal payment due date and go on till 180 days after the due date. On an average, each policy holder is given 9–10 reminder calls during the period of 180 days, starting the policy due date. Efforts are stopped for the cycle, when the customer pays the renewal premium. If the customer does not pay till 180 days after the due date, the policy is considered to be *lapsed*. Figure 3 provides an overview of activities performed for policy renewal payment collections.

In the absence of any analytical model, every policy holder was reminded for renewal payments.

2 Source of Data

All the data used in the study is of Max Life Insurance except some macroeconomic data published on Indian government Web sites. Various *workshops were held between the analytics team, customer retention team and the IT team* to identify the relevant variables. A *map of customer journey with Max Life was prepared* to identify customer touch points and map the relevant data available with logical considerations for the data points impacting lapsation rates. Figure 4 lists the different types of data used for the research.

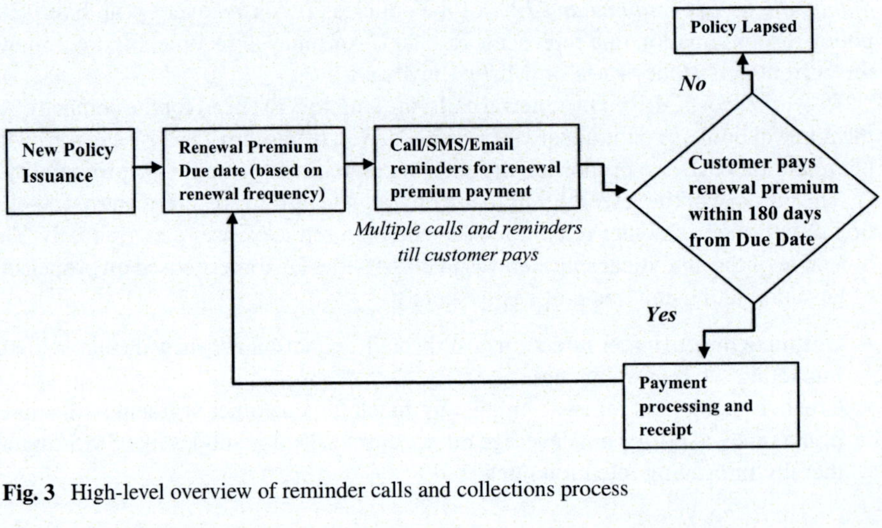

Fig. 3 High-level overview of reminder calls and collections process

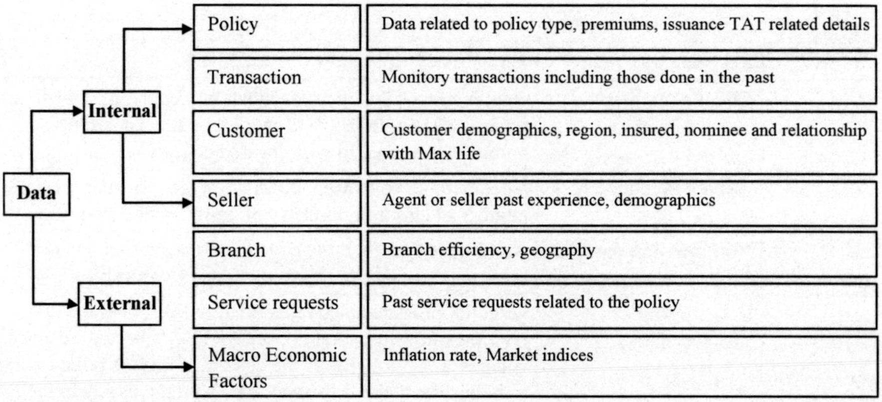

Fig. 4 Types of datasets used

2.1 Description of Datasets

2.1.1 Policy Data

The policy-related variables were considered to understand the journey of the policy from issuance, nature of preference from the customer in terms of type of policy, its alignment from the customer demographics and needs. Policy premium-related variables indicate amount of investment done as compared with the income impacting commitment for policy continuation. Past due payments and policy-related transactions were considered to understand the commitment toward policy continuation (Table 1).

Table 1 Policy-related variables

S. No.	Subcategory	Variable descriptions
1	Issuance	TAT for issuance, underwriting comments, issue/due month, current age from issuance, medical examination done if any, sales channel
2	Type	Premium payment term, sum assured, riders, coverage, renewal mode (yearly, monthly, etc.)
3	Premium	Annualized premium amount, payment mode, premium paid/outstanding
4	Past transactions	Premium payment patterns (on time, late, early, etc.), if policy was recovered, lapsed in the past, payment mode changes, fund value (in case of ULIP products)

Table 2 Customer-related variables

S. No.	Subcategory	Variable descriptions
1	Client demographic	Age, income, education, occupation, gender, marital status, no. of children, industry of work, client occupation, nominee relationship with insured client
2	Insured demographics	Age, income, education, occupation, gender, marital status, number of children, industry of work, smoker flag
3	Nominee demographic	Nominee—age, income, education, occupation, gender, marital status, no. of children, industry of work
4	Relationship	Total no. of issued, active, applied polices, client active relationship value, total no. of "ARP"—"aborted, rejected and pending" policies of the customer, orphan policy—flag to indicate whether the agent who sold the policy to the customer is still with Max Life
5	Geography	Client permanent address pin code, client current address pin code

2.1.2 Customer Data

The customer-related variables consisted of mainly the customer, insured and nominee demographics and their relationship with Max Life, indicated by the number of policies bought, amount of investment done with the company etc. (Table 2).

2.1.3 Seller or Agent Data

Agents and distribution partners play an important role in the quality of sales done, and quality of sales is an important contributor to ***lapsation***. Max Life has different types of sales channels including a network of Max Life agents and distribution partner banks. The seller- and agent-related variables captured information such as agent or seller demographics, their experience in sales and quality of sales done, volume of business done for Max Life (Table 3).

2.1.4 Selling and Relationship Branch Data

Sales branches play a significantly important role in both getting new customers and maintaining relationships with existing customers. Branch performance metrics such as *branch efficiency* and past persistency trends for policies sold from branches become important indicators in understanding lapse rates. Geography as an important indicator of quality of life and well-being of customers is a determinant to customer interest and awareness about the life insurance products. Geography of branch, hence, was hypothesized to play an important role in determining lapsation (Table 4).

Table 3 Agent-/seller-related variables

S. No.	Subcategory	Variable descriptions
1	Demographic	Agent age, gender, education, marital status, agent residence postal code, vintage with Max Life
2	Experience	Years of experience, evaluation score of the agent while hiring on 5-pointer scale, active/suspended status of Max Life agent as of the month, flag to determine whether sales agent was reinstated (after termination) any point of time in his association with Max Life, persistency of policies sold by the agent
3	Book (business done for Max Life)	Total count/value of policies, issued by the sales agent, total number of policies, issued by the sales agent, where client died, proportion of active, policies with claims, applied to issued proportion

Table 4 Branch-related variables

S. No.	Subcategory	Variable descriptions
1	Book (business done for Max Life)	Count and value of policies applied, issued, active by the office/go since inception, number of policies with early claims, deaths, etc.
2	Efficiency	13 M branch persistency—ratio of paid premium to collectible/due premium (for 13-month due policies) at branch level for the month
3	Geography	Selling branch age, selling branch zone/city

2.1.5 Product Data

Data related to product consisted of variables such as type of product, for example if the product was *Traditional Life Insurance* plan or *Unit Linked Insurance Plan (ULIP)* or *Online Term plan.* Type of plan selected by a customer is an important indicator of alignment of customer need with the chosen plan. Apart from that, some plans are very popular and have better market performance. Such differential performance across plans was expected to be an important driver of lapsation (Table 5).

2.1.6 Service Request

The service request data consisted of *dispositions* from past touch points with the customers. These variables play a very important role in understanding lapse

Table 5 Product related variables

S. No.	Subcategory	Variable descriptions
1	Type	Name of the product/plan of the policy, plan vintage
2	Performance	Product sales (value and volume), issued, active, persistency, recovered policies, aggregated at plan (product level)

Table 6 Service request-related variables

S. No.	Subcategory	Variable descriptions
1	Contact	Total number of service requests received from the customer, any service request open at the time of policy due
2	Sentiment	Customer sentiment after the solution provided by the service agent—negative, positive, neutral, total no. of service requests pertaining to surrender of the policy

rates. Intuitively, dissatisfied customers have negative dispositions and are likely to discontinue the policy and maybe difficult to retain (Table 6).

2.1.7 Macroeconomic Variables

A number of studies have been published about lapsation modeling, highlighting the impact of macroeconomic conditions on the life insurance industry. Taking an inspiration from findings of such findings, a study of 6 macroeconomic variables was studied, as possible determinants of lapsation. With a significant correlation observed among these variables, 3 variables, gross domestic product (GDP), short-term interest rates and Nifty Index, were used after appropriate transformations.

3 Methodology

The overall methodology involved applying the research to build a solution which can be used to solve the business problem at hand. A 7-step process was followed to build a deployable solution. The solution was used to classify renewal policies on the basis of propensity to ***lapse***. Figure 5 shows the seven steps followed for the research methodology.

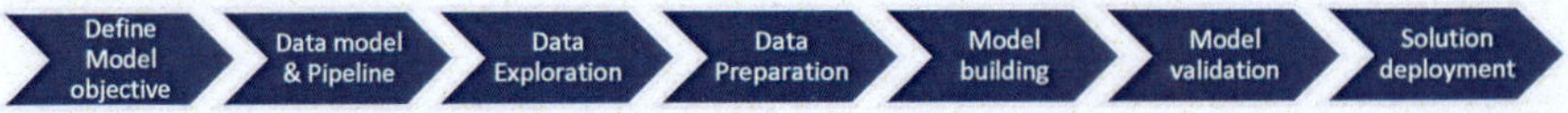

Fig. 5 Overall methodology of the research

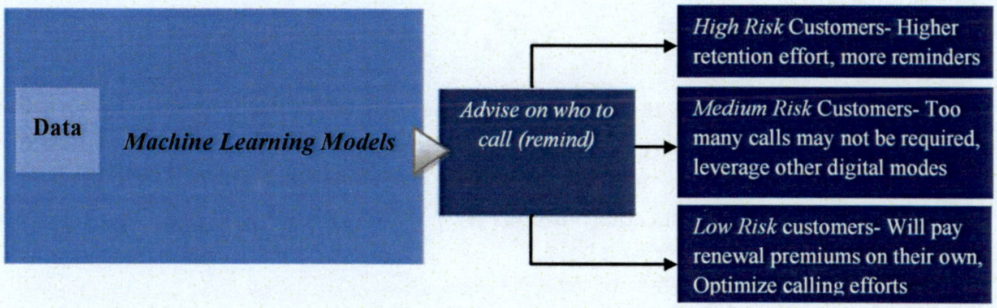

Fig. 6 Overview of how the solution is expected to solve the business problem

3.1 Defining Research Objective

Following the business objective, the research aimed at optimizing the customer retention costs and improving the renewal collections, through executing a differential retention calling strategy for renewal policy book. This differential strategy would be based on classification of policies on propensity to lapse. The research objective was to build an advanced analytics model to determine the probability of a policy to lapse within 180 days from the due date of renewal payment. Once the lapsation probability scores could be obtained, policies would be clubbed to define 3 key segments, *high risk*, *medium risk* and *low risk* of lapsation. From the strategy execution point of view, *high-risk* customers would be given more and frequent reminders to maximize collection. The *medium-risk* customers can be reached out by digital means such as *SMS* and *emails*, which are less expensive than telephonic calls. The number of renewal calls for the Low Risk customers can be minimized as they represent customers who require less intervention to pay the renewal premiums. Figure 6 provides an illustrative view of end state as envisioned from the output of the research.

3.2 Data Extraction

Max Life has a wide landscape of technology systems used for different business processes across the value chain. The data required for the research resided in 8–10 different systems, and as a first step, a data model and pipeline were required to be built by collating all the required data at one place. The required datasets were combined, and first level of processing was done to create a centralized database. This was done using ***Informatica*** and ***Base SAS (Statistical Analysis System)***. The processed dataset was then used for performing the next steps.

3.3 Data Exploration

Exploratory data analysis was carried out as a first step to prepare the data. Descriptive analysis was done to identify distributions and isolate missing values and outliers. A few numerical variables were found to be skewed, because of which variable transformation was done to reduce the level of skewness.

3.4 Data Processing

A number of data processing steps were required, before the data can be used for building the analytical models. One of the most important considerations was the form of the data. Since 3.5 years of past renewal data was used, there were multiple *due dates* for same policies present in the data. For example, an *annual* renewal mode policy will have *due dates* every year and hence the policy will appear 3 or 4 times in the entire dataset. Hence, *policy number*, which is a unique identifier for the policy, could not be used directly as unique identifier. *Policy number* along with the *due date* for renewal payment was used as the unique identifier for each row of the dataset. Data processing was a 5-step process, as illustrated in Fig. 7.

3.4.1 Outlier Treatment

A few variables, such as client income, were found to have some outliers, and the outlier treatment was done using flooring and capping techniques.

3.4.2 Missing Value Imputations

Since the data was coming directly from the source systems, some variables had missing values which were required to be cleaned. Variables such as *agent education* and *client industry*, which had more than 20% missing values, were excluded from the dataset. In some cases, the absence of values could not be considered as a missing value. One example is *nominee* demographics. If the policy has no nominee, then the nominee details were not present for that policy. As such variables cannot be directly used in the model, these variables were marked using a flag indicating if

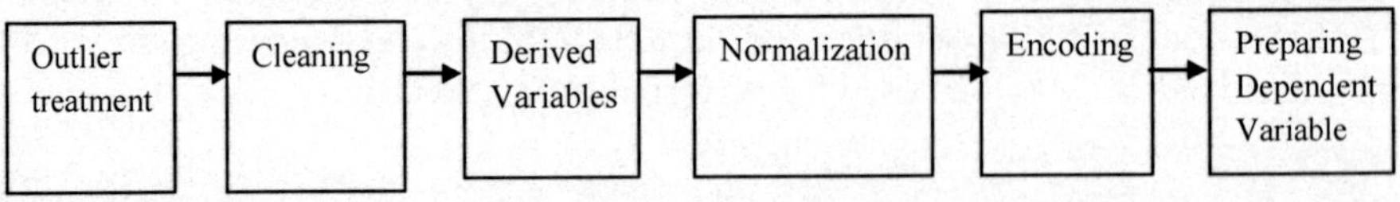

Fig. 7 Overall data processing steps used during the research

a nominee is present for a policy or not, and these flags were used instead. Other important variables such as *client income* were found missing for 8% of policies, and these were imputed using *proc MI*, with appropriate parameters, which uses different algorithms to impute missing values.

3.4.3 Derived Variables

Apart from the raw variables, a number of derived variables were also calculated to increase the explainability of *lapsation* behavior. Although, theoretically, derived variables and variable reduction techniques are not required for *deep learning* models, since we also build *supervised machine learning* models such as *logistic regression*, some derived variables were created. Largely, these derived variables consisted of ratios for direct variables. For example, instead of considering number of policies issued by a *branch* and number of policies active from that *branch*, we considered a ratio of these two variables to arrive at a relative measure, which are comparable across branches. There were 20 derived variables prepared to be used in the models.

3.4.4 Variable Reduction

Although variable reduction techniques are not considered to be required by *deep learning* models, assisting the models with an input on most relevant variables helps improve the model performance. There were 150 variables as picked up for the research, before analyzing the impact of each independent variable on lapsation. For all the variables, bivariate plots showing lapse percentage at various levels of independent variables were studied to understand the impact of varying levels of variable on event (lapsed policies). Figure 8 shows the bivariate analysis of lapsation across sales channels. It can be observed that there is a noticeable difference of lapse rates across channels; hence, channel as an independent variable would be important to use in the models.

Along with the bivariate analysis, *information value*, using the *weight of evidence method*, was calculated for each variable. Independent variables with information value above 0.05 were considered to be important for explaining lapse characteristics.

For the purpose of testing *supervised machine learning* models like logistic regression, multicollinearity was checked using *variance inflation factor* obtained from SAS. Variables with VIF greater than 7 were dropped from the supervised learning models. After the application of all the above techniques, a final set of 61 variables was used in the final modes.

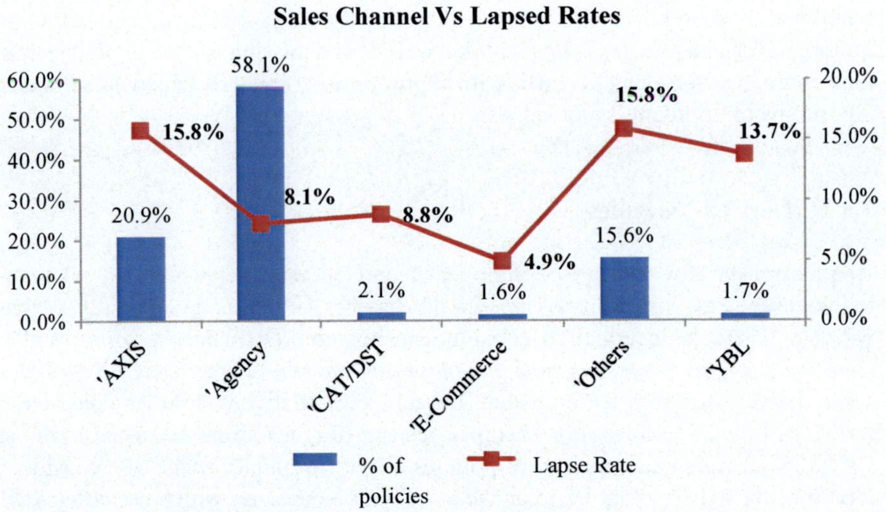

Fig. 8 Bivariate plot of lapse rates versus sales channel

3.4.5 Normalization

As a part of data processing, continuous independent variables such as *premium amount*, *sum assured and branch persistency* were normalized by using a *min–max normalization methodology*, with the scaling done in the range of 0 and 1. *Normalization* of continuous variables is required to bring all continuous variables to a common measurement scale.

3.4.6 Encoding

The *categorical* independent variables were converted to *binary* values, before they could be used in the model. This process is called as *one-hot encoding*. While it converts the categorical variables to numeric, it also ensures that individual effect of specific categories is captured independently, rather than the combined effect of that categorical variable.

3.4.7 Preparing Dependent Variable

The dependent variable considered was a binary variable indicating whether a policy *lapsed* (1) or not (0). A policy is considered *lapsed* if the customer does not pay the renewal premium amount till 180 days from the payment due date. There are different cases of *lapse* based on reason for discontinuation of the policy. Figure 9

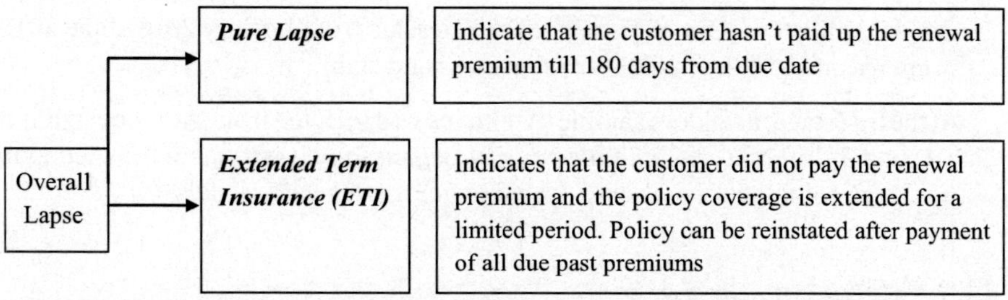

Fig. 9 Subcategorization of lapsed policies as considered for this research

defines two different cases of lapsation considered for the purpose of building the models.

Effectively for the company, both these types of lapses are considered as policy discontinuation because there is no future renewal income expected from these policies. Both ***pure lapse*** and ***ETI*** have a 50–50% contribution to overall lapse events.

The customer retention operations focus on both ***pure lapse*** and ***ETI*** categories. The key reason for this is that the policy can be reactivated from these 2 states (pure lapse and ETI) and proactive retention efforts through calls, SMS and other digital channels help controlling such events.

In order to test out the predictive power of the predictive power of the models, only ***pure lapse*** events were considered. Pure lapsation is a larger problem, and proactive efforts are expected to make higher (than ETI) impact on retaining customers; hence, being able to predict ***pure lapse*** policies adds the highest business value.

3.5 Model Building

An iterative three-stage model development process was followed. The first stage comprised developing models using different methodologies on a representative sample and dropping the models showing lesser predictive capabilities. In the second stage, final selected models were created on full dataset. The third step involved testing of the developed model for stability and robustness.

Before the model building exercise, the final dataset was split into 3 parts.

1. Model Building Data: All policies *due* from January 2015 to December 2017 were studied and further split into parts.
    a. Training Data: A sample comprising 70% of policies is selected at random to create and train the model.
    b. Validation Data (Out of Sample): The remaining 30% of policies is used for validation.

2. Model Validation (Out of Time): All policies due from January 2018–June 2018* were used to test the model for robustness and stability.

At the first stage, a random sample of policies was selected from the model building dataset and 3 different *machine learning classification* algorithms were tried along with various combinations of *deep learning neural network* models.

3.5.1 Logistic Regression

A binary classification model was built using classic logistic regression technique to classify ***pure lapse*** events across the renewal policy book sample.

3.5.2 XG Boost

Extreme gradient boosting, commonly known as *XG boost,* has become a widely used and really popular tool among data scientists in industry. A gradient boost model was built with the same sample dataset with *pure lapse* as dependent event.

3.6 Deep Learning Model (Neural Networks)

For building a deep learning model to predict pure lapse events, various combinations of feedforward neural network (FNN) models were tried and the final deep learning model was iteratively crafted using a combination of feedforward neural network (FNN) and long short-term memory (LSTM) neural network architectures. The LSTM layer was built specifically to use the past interactions of the customers, such as payment transactions and service request data. The final architecture of the DL model is shown in Fig. 10.

3.6.1 Model Performance

For the purpose of assessing the predictive power of the models, the *cumulative lift* or ***capture rate*** was considered to be the most important metric. *Cumulative lift* shows the ability of the model to separate out the events and non-events. It is measured by *cumulative sum of percentage of events* captured in each decile. It is also referred to as the *capture rate* (indicating the events captured in each decile and cumulatively for 10 deciles).

Capture Rate (or Lift)= Total events in a decile / Total number of events in the dataset

Cumulative Capture Rate= Cumulative sum of events captured, for deciles

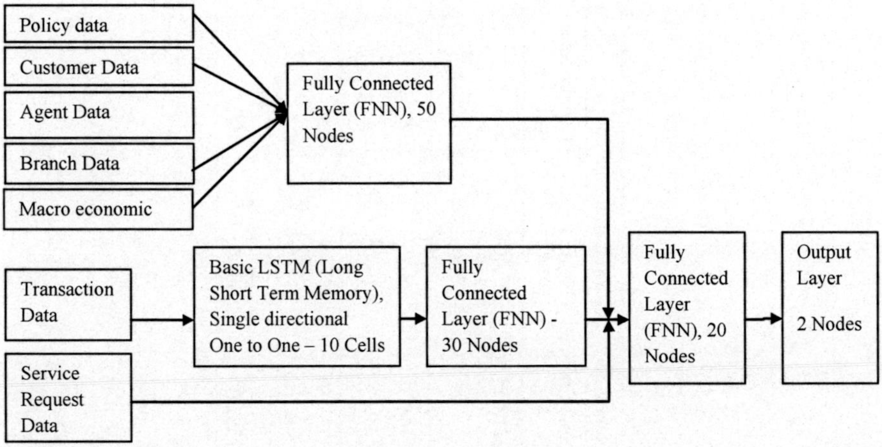

Fig. 10 Deep learning neural network model architecture

In the context of the business objective, a higher *capture rate* in *top deciles* indicates higher concentration of *pure lapsers* in the *high-risk* bucket and higher confidence to concentrate retention efforts on top deciles; at the same time, a low *capture rate* in the lower deciles indicates a lower risk of any potential loss of *renewal income* while optimizing the retention efforts in the *low-risk* bucket.

4 Results

For assessing the predictive power of the models, three classification techniques were tried, with a dataset containing 61 finalized variables, and the *capture rate* for each of these model outputs was compared. Table 7 provides a comparison of model performance across the 3 techniques.

The logistic regression model sharply distinguished the *pure lapse* events capturing *84% events in the top 3 deciles (top 30% population)*. However, for the case at hand, *XG boost* algorithm performed poorly as compared with *logistic regression* model probably because this technique is more suited for ensemble model outputs rather than a stand-alone classifier. An overall cummulative capture rate of *75% was achieved in top 3 deciles*, after multiple iterations.

The *neural network*-based *deep learning* model turned out to have the highest predictive power. Not only the model captured *89% pure lapsers in the top 3 deciles*, it also reduced the *capture rate to 0% in the bottom 3 deciles*. The long short-term memory (LSTM) network effectively captured the interaction behavior of customers based on past premium payments and service call sentiments.

Following the superior performance of the *deep learning* model, it was built on the complete dataset with 3 lakh policies (some policies had to be removed from

Table 7 Comparison of capture rates across 3 machine learning models

Risk bucket	XG boost			Logistic regression			Neural network deep learning model		
	No. of policies	No. of events	Capture rate (%)	No. of policies	No. of events	Capture rate (%)	No. of policies	No. of events	Capture rate (%)
High risk (deciles 10, 9 ,8)	221,653	23,147	75	221,653	25,924	84	221,653	27,468	89
Medium risk (deciles 7, 6, 5, 4)	312,201	6172	20	312,201	3703	12	312,201	3395	11
Low risk (deciles 3, 2, 1)	221,450	1543	5	221,450	1235	4	221,450	0	0
Total	755,304	30,862	100	755,304	30,862	100	755,304	30,862	100

the dataset while data cleaning and processing). The architecture of the DL model was maintained as is for the larger dataset, with a train and test split of 70% and 30%, respectively. The train and test model performance indicated a more than *99% capture rate in the top 3 deciles*. Although this seemed to be an overfitted model, however the out-of-time sample performance of the model turned out to be 90%, which was seen to be stable, with a 3% (percentage points) variance in *capture rates*, over the months across January'18 to June'18. Tables 8 and 9 provide a decile-wise *capture rates* for the final *deep learning* model.

Since proactive efforts for retention play an effective role in case of *pure lapse* and *ETI* policy discontinuation, the chosen *deep learning* architecture was also applied

Table 8 Decile-wise capture rates for train and test dataset (January'15–December'17 due policies)

Decile	Number of policies	Number of events	Percentage events captured (%)	Cumulative capture rate (%)	Risk tag
10	305,765	192,634	84	84	High
9	305,765	34,229	15	99	
8	305,765	2378	1	100	
7	305,765	11	0	100	Medium
6	305,765	1	0	100	
5	305,765	1	0	100	
4	305,765	2	0	100	
3	305,765	0	0	100	Low
2	305,765	0	0	100	
1	305,765	0	0	100	
	3,057,650	229,256			

Table 9 Decile-wise capture rates for out-of-time dataset (January'18–June'18 due policies)

Decile	Number of policies	Number of events	Percentage events captured (%)	Cumulative capture rate (%)	Risk tag
10	202,047	48,587	45	45	High
9	202,047	31,314	30	75	
8	202,049	15,911	15	90	
7	202,048	10,191	10	99	Medium
6	202,050	983	1	100	
5	202,046	16	0	100	
4	202,049	6	0	100	
3	202,048	0	0	100	Low
2	202,047	4	0	100	
1	202,048	2	0	100	
	2,020,479	107,014			

Table 10 Capture rates for deep learning models with different types of lapse events

Model	Model event (1)	Train		Test		Out of time	
		Top 3 deciles (high risk) (%)	Bottom 3 deciles (low risk) (%)	Top 3 deciles (high risk) (%)	Bottom 3 deciles (low risk) (%)	Top 3 deciles (high risk) (%)	Bottom 3 deciles (low risk) (%)
Model 1	Pure lapse	99	0	99	0	90	0
Model 2	ETI	88	0.15	88	0	84	0
Model 3	Combination of pure lapse and ETI	81	2	81	2	72	3

to iteratively create models for ETI events and a combination of both types of *lapse events* (*pure lapse and ETI*).

It can be observed that *deep learning* model performs quite well for modeling *ETI* events, as well, with a *capture rate of 87% and 84% in top 3 deciles, for train, test and out-of-time samples, respectively*; however, a combined model with both ETI and pure lapse (event was defined as either ETI or pure lapse = 1) drops significantly indicating there might be some difference between nature of policies which move to *pure lapse* as compared with those which move to *ETI* (Table 10).

5 Conclusion

The deep learning model, with the designed architecture, came out to be the most superior in terms of capturing both *pure lapsers* and *ETI* events. Both *in-time* and *out-of-time* validations were done, and the model is robust and stable over time and does not witness any significant drop in ability to predict the lapsers.

Although predictor explainability is difficult to be derived in a deep learning model, a both information value and variable dropping exercise were conducted to get a sense of important predictors of lapsation. A list of important predictor variables was compiled, and Table 11 presents the description of these important variables.

The final results showed that a total of *17 variables were important to predict lapsers* at the time of policy due date. The final model capturing *pure lapsers* was selected to be deployed, and appropriate decile-wise risk tags were created to use the model outputs for customer retention strategy. As a further research, the combined model for pure lapse and ETI is being refined further to see if both these types of events can be modeled together using ensembling techniques.

Table 11 Description of important variables

Variable category	Variables
Policy	Policy vintage, premium amount, channel, billing mode, coverage term, medical/non-medical issuance, due month (seasonality factor), policy sold through corporate customer, policy recovered in the past
Agent	Agent vintage, agent past lapse rate
Product	Plan family (type of plan)
Seller branch	Branch persistency
Customer	Age, occupation, income, gender, residential address zone
Transaction	Timely payer/late payer
Service requests	Negative sentiment of service request calls

6 Implications

Following the purpose and objective of the research, the outcomes were used to create an AI/deep learning-based predictive intelligence solution for the customer retention strategy.

An end-to-end integrated solution was built to classify customers based on risk of lapsation, using the *deep learning* model. The risk classification is used to devise a differential retention calling strategy. The solution also tracks retention efforts with overall cost and renewal income monitoring.

The solution enables execution of a differential retention strategy based on risk classification by

- Proactively reaching out to *high-risk* (high chance lapsation) customer segment to educate them on policy benefits, thereby increasing customer retention.
- Reducing the number of renewal calls made to *low-risk* customers and leverage other communication modes like SMS and digital.

We estimate that, using this solution, *Max Life will be able to save 12–15% customer retention cost without any drop in renewal income.* In the long term, we will be able to use the solution for proactive retention efforts in *high-risk* segment to increase retention rates, thereby increasing the renewal income (*with goal of 1% improvement in retention rates, thereby adding Rs 100 crore to renewal income*).

Apart from the quantitative benefits, the research-based solution has already achieved a number of qualitative benefits including

- *Business integration* with cutting-edge, sharp data science algorithm.
- *New business quality improvement* driven by customer segment characteristics which define risky customers.
- *Process and metrics standardization* achieved during data preparation from various different source systems.

References

insure@nceiknowledge admln experts speak. (2018, 7 July). Technology can help solve one of the largest problems in Life Insurance in India: The Persistency Ratio. https://www.aegonlife.com/insurance-investment-knowledge/life-insurance-india-persistency-ratio/.

Jones, T. L. (2017, April). Indian consumers positive to using mobile and apps for-insurance. https://blogs.lexisnexis.com/insurance-insights/2017/04/indian-consumers-positive-to-using-mobile-and-apps-for-insurance/.

Optimization of Initial Credit Limit Using Comprehensive Customer Features

Shreya Piplani and Geetika Bansal

1 Introduction and Motivation

With a total of 36.24 million credit cards in operation with a spend of Rs. 41,437 crores in January'18 from a 28.85 million credit cards and usage of Rs. 32,691 crores in January'17, credit card market has shown an incredible growth in India. During the initial introduction of credit cards in the Indian market, the word credit did not go along with the Indian mentality, believing that credit cards would increase their liability and might lead to payment of huge interests, if not cleared on time. The tremendous growth in the recent times of this market can be accredited to the acceptance of 'spend now and pay later' strategy which was supported by the ease of digital payments, acceptance in almost every monetary transaction and the e-commerce boom along with the option to repay in easy instalments.

Traditionally, credit card issuers have relied upon the income and the score of an applicant to calculate his credit limit. In the current scenario, the average utilization on credit cards is only 30%. This highlights that a majority of customers have a huge unutilized limit, resulting in capital blockage for the institution. Additionally, post-issuance, 23.3% of the applicants do not activate the card. Hence, it is necessary to devise a dynamic and reliable method to determine credit limit which focuses on crucial factors like expected spending, odds of activation given the limit, behaviour on similar account, etc. This paper proposes a more granular methodology to determine the credit limit of a customer based on his expenditure potential and his credibility to payback based on his credit history, payment history and various demographic variables. The idea is to understand the role of the above factors and the methodologies adopted to address the problem of limit assignment.

S. Piplani (✉) · G. Bansal
Experian Credit Information Company of India, Mumbai, India
e-mail: shreya.piplani@experian.com

G. Bansal
e-mail: geetika.bansal@experian.com

A. K. Laha (ed.), *Applied Advanced Analytics*, Springer Proceedings in Business and Economics, https://doi.org/10.1007/978-981-33-6656-5_7

2 Literature Review

Businesses granting credit through credit cards faces numerous challenges with the growing demand and varied consumer behaviour. Researchers over the time have focused on credit limit increase and decrease post observing payment patterns on the account over a specified time period. Questions on usage of credit line, payment patterns in terms of revolving, profitability of the customer and how likely is the customer going to default in future have been raised and discussed. Various segmentation and prediction techniques have been devised and tested. Bierman and Hausman (1970) formulated a dynamic programming model in which the decision process focused on whether to grant credit or not and for what amount. In their formulation, the amount of credit offered was linked to the probability of non-payment or default. Haimowitz and Schwarz (1997) developed a framework of optimization based on clustering and prediction. Expected net present value is calculated at multiple credit lines combined with the probability of cluster memberships. This paper also highlights the future scope in terms of using multiple independent variables for optimization and studying their effects using other techniques such as neural networks. Research by Hand and Blunt (2001) highlighted the use of data mining techniques to predict spending patterns in a database of UK credit card transactions, specifically on the petrol stations. Dey (2010) highlighted the importance of understanding the action effect models and addressed the prominence of each component of revenue and risk. The research by Terblanche and Rey (2014) focused on the problem of determining optimal price to be quoted to the customer, such as income of the lender is maximized while taking price sensitivity into account. Using probability of default, loss given default and other factors, an equation describing the net present income is developed. Budd and Taylor (2015) presented a model to derive profitability from a credit card assuming that the card holder pays the full outstanding balance. Most of the research either focused on one constituent of optimal allocation or followed a single approach that is either from a pure risk or revenue perspective. Even though the elements described in this paper are discussed in multiple researches either one or more of the approaches, methodology and techniques are missing.

3 Methodology

The methodology to obtain an optimal initial credit limit for each customer is based on the information submitted during the application and bureau history of the customer (in case he/she has a previous line of credit). In this context, allocation and maintenance of pertinent credit limit will maximize revenue and minimize the risk associated.

Understanding the revenue and risk components, theoretically, a high credit limit has an advantageous effect of increased expected revenue, but at the same time both the probability of default and expected exposure at default also increase. Similarly, a low credit limit decreases expected loss but leads to a decrease in expected revenue. The overall effect and the appropriate action depend on which of these effects is stronger (Dey 2010).

Revenue generated from any credit card portfolio is influenced by complex interactions between several factors like probability of activation, probability of attrition, propensity to revolve and credit limit utilization, while the risk component comprises the probability of default at the time of acquisition, behavioural probability of default and exposure at default (Dey 2010).

The behaviour of each effect variable is modelled separately, and their combined effect defined as risk-adjusted return is studied.

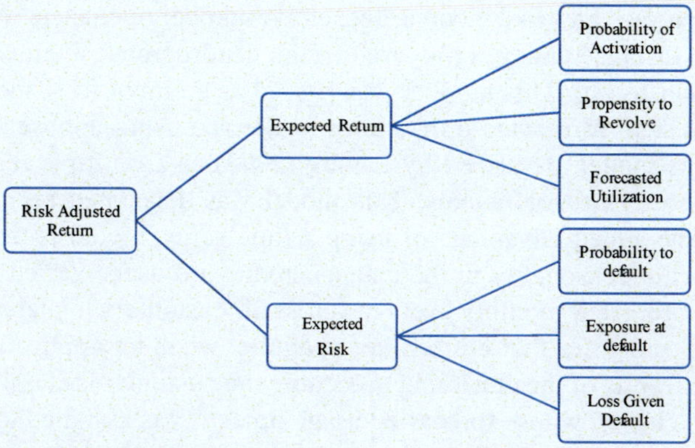

Additionally, for modelling the above components monthly payment data, comprising statement data, balances, minimum amount due and the payment date of credit card customers were used.

- **Probability of Activation**: This model results in the probability of activation of the card given the limit assigned. The dependent variable was calculated using application data of customers. The value 1 was assigned to customers who activated the card, given a specific limit and 0 to others. Multiple (predictive) statistical techniques like logistic regression, gradient boosting and random forests were used to derive the predicted probability of activation. The models were then compared based on misclassification rate to obtain the final model.
- **Probability to Revolve**: One of the key constituents of the revenue component is interest, which is calculated on the deferred payments. For example, a customer pays a certain percentage of the current balance each month. The rest of the unpaid amount incurs an interest. For predicting this factor, customers are classified into 4 categories:
  - Transactor: A customer who pays the exact balance due each month and hence does not incur any interest charges
  - Accidental Revolver: A customer with deferred payment for less than or equal to two months

- Acute Revolver: Revolved less than or equal to 5 times
- Chronic regular revolver with more than 5 months of deferred payment.

Multinomial regression was used for calculating the odds of a customer belonging to a particular segment. Another model with a binary-dependent variable for revolver/non-revolver was developed, and the two models were compared using the Gini coefficient.

- **Forecasted Utilization**: Credit card customers fall into widely diverse categories; hence as a first step, the customers were clustered based on various behavioural variables like utilization of credit limit, delinquency on credit card, maximum delinquency on other accounts, payment patterns on credit card and additionally application variables like income, age, etc. A number of clusters were determined using the distance changes observed in the dendrogram, obtained through hierarchical clustering. This number was passed as an input to K-means clustering. As a next step, forecasted utilization was derived using unobserved component time series model, described by Famby (2008) as a multiple regression model with time-varying coefficients. This model was developed separately for each cluster. The added advantage of using a time series model is that incorporated the impact of seasonality in the data as credit card usage varied from season to season. Forecasted monthly average across all customers in a cluster formed the usage trajectory for that cluster. For example, when an application is received, using the results of the clustering procedure the customer is classified as a part of a specific cluster, whose forecasted usage trajectory is assigned to that customer.
- **Probability of Default**: This parameter is defined as risk for the institution at the time of acquiring the customer. If from the date of application, in the next 12 months, any customer with a delay past the due date of greater than or equal to 90 days was classified as a bad customer and good otherwise. Logistic regression was used to estimate the value of this parameter.
- **Exposure at Default**: This parameter is calculated based on the assumption that an account is likely to go bad, when utilization is maximum. The maximum forecasted utilization and the behavioural probability of default were used to calculate the value of this factor.
- **Behavioural Probability of Default**: Post acquiring, the likelihood of a customer falling in a particular default bucket in future based on the observed behaviour is the value for this factor. For this purpose, we have used the bureau score that comprehensively captures the performance of a customer over existing accounts.
- **Loss Given Default**: For this analysis, loss given default is assumed to be 100%. This implies that if the customer defaults on the portfolio, total outstanding balance is defined as the total loss.

The function of the above elements was used to optimize the initial credit limit of the customer.

RAR = *f*(*Expected revenue, Expected risk*) where:

f (*expected revenue*) = *f*(*expected utilization, probability of activation, probability to revolve, interest rate, interchange*) *f* (*expected* risk) = *f(probability of default at acquisition, exposure at default, loss given default, behavioural probability of default)*

The difference between the two, defined as the risk-adjusted returns, is maximized to obtain the optimal limit. The constraints for this optimization problem are linear combination of borrower's behavioural characteristics (affordability, etc.), lender's risk constraints (risk appetite, exposure, etc.) and operational requirements of the business (acquisition book size, costs, etc.). Initial convergence to the objective function maxima was attained through the most commonly used method known as the Lagrange multipliers. In addition, since the objective function and constraints combined lead to a convex optimization problem, Frank–Wolfe algorithm is used to attain the optimal limit corresponding to maximum RAR. The Frank–Wolfe method solves one subproblem at each iteration, produces a sequence of viable solutions over the region of interest and hence is computationally feasible (Freud and Grigas 2014).

4 Benefits to the Business

A method of modelling and determining the initial credit limit consistent with the objective of maximizing revenue and minimizing risk has been discussed. Each component being predicted can be applied separately to most customer decisions across the customer life cycle: customer acquisition and customer account management. Businesses based on their requirement can understand the trade-off between different scenarios, thereby enabling them to determine the best action for each customer. The approach enables improved customer decision-making process in terms of:

- Moving away from the traditional methodologies to taking a holistic view of the customer actions and decisions in terms of his spend and repayment behaviour
- Helping the business to understand optimal capital allocation through credit card
- Helping the business identify homogenous customer segment and design targeting strategies accordingly
- Using individual component models, to understand characteristics of customer segments, devise varied strategies and schemes to enhance customer experience and helping the business reduce churn and ensure adequate customer loyalty.

References

Bierman, H., & Hausman, W. (1970). The Credit Granting Decision. *Management Science, 16*(8), B-519–B-532.

Budd, J., & Taylor, P. (2015). Calculating optimal limits for transacting credit.

Dey, S. (2010). Credit limit management using action-effect models.

Famby, T. (2008). The unobservable components model.
Freud, M. R., & Grigas, P. (2014). New analysis and results for the Frank-Wolfe method.
Haimowitz, I. J., & Schwarz, H. (1997). Clustering and prediction for credit line optimization.
Hand, D. J., & Blunt, G. (2001). Prospecting for gems in credit card data. *IMA Journal of Management Mathematics, 12*(2), 173–200.
Leippold, M., Vanini, P., & Ebnoether, S. (n.d.). Optimal credit limit management under different information regimes.
Terblanche, S., & Rey, T. D. (2014). Credit price optimisation within retail banking.
Verma, S. (n.d.). MediaNama. Retrieved from https://www.medianama.com.

Mitigating Agricultural Lending Risk: An Advanced Analytical Approach

Aditi Singh and Nishtha Jain

1 Introduction

As per the Situation Assessment Survey (SAS) for Agricultural Households by NSSO 70th round, in 2012–13, almost 40% of the agricultural households still relied on non-institutional sources for their credit needs, an increase of almost 11% over 1990–91. Moneylenders form a major part, around 26%, of that non-institutional credit. Even with the rising credit disbursements and loan waivers, we have not been able to improve the situation of our farmers. Empirical and situational evidences suggest that generalized loan waivers have made less than marginal contribution toward improving credit situation of farmers (FE Online 2018). It rather creates a situation of moral hazard which affects the loan repayment behavior of all the farmers. Since 2011–12, percentage of bad loans from agriculture sector has climbed every year and the growth rate of loans disbursed to this sector has become close to stagnant. In FY 2018, banks disbursed only an additional 6.37% to this sector which is the lowest in a decade (Iyer 2019).

Agriculture sector poses risks for the banks in multiple forms. Lending to the agriculture sector has been adversely affected in recent times, and it could be indicative of the deteriorating asset quality (Trends and Progress Report, RBI 2017–18).

Reserve Bank of India (RBI), which is India's central bank and regulating body, oversees the functioning of all the banks that operate in the country and has identified certain priority sectors, of which agriculture is also a part, to ensure necessary credit flow to these sectors. However, banks especially the private and foreign banks are not familiar with India's agricultural landscape and feel reluctant to lend in this sector (Jayakumar 2018). Because of inadequate knowledge of the risks pertaining to these sectors, they refrain from direct lending and instead end up investing in Rural Infrastructure Development Fund (RIDF) of NABARD or buying Priority

A. Singh (✉) · N. Jain
Data Modeler, Experian Credit Information Company of India, Mumbai, India
e-mail: aditi.s11@gmail.com

A. K. Laha (ed.), *Applied Advanced Analytics*, Springer Proceedings in Business and Economics, https://doi.org/10.1007/978-981-33-6656-5_8

Sector Lending Certificates (PSLCs) to meet their priority sector lending targets. This paper aims to shed some light on the financial landscape of agricultural sector in India to help banks understand the market and model the associated credit risk in an improved manner and hence bridge the gap between the borrower and the lender.

Credit risk associated with an individual can be classified into two broad categories:

1. Capacity to pay and
2. Intention to pay.

Capacity to pay is governed by the principle that the individual should have the ability to generate a steady flow of income which depends largely on his demographic features such as age, qualification and profession. These features along with income and existing assets of the individual determine whether the individual has the capacity to pay back the loan. Following Maurer (2014), risks in agriculture finance can be broadly classified into the following categories which influence an agricultural household's capacity to pay:

i. Production Risks: Agriculture production in India is fraught with the risk of poor monsoon, disease and pests due to which farmer's income suffers. Lack of proper irrigation facilities, immense dependency on monsoon, lack of good quality seeds and chemical fertilizers can lead to suboptimal output and therefore insufficient generation of income to pay back the loans.
ii. Market Risks: There are price uncertainty and volatility associated with farming where farmers do not know at the time of plantation what prices their produce would fetch. The interplay of demand and supply factors in determining market prices causes agriculture income to be volatile. Minimum support price (MSP) plays a crucial role here in defining a floor price at which government would procure crops from farmers. The Cabinet Committee on Economic Affairs (CCEA), Government of India, determines the MSP based on the recommendations of the Commission for Agricultural Cost and Prices (CACP). The objective of MSP is to protect farmers from the price shocks and to ensure food security through buffer stocks and Public Distribution System (PDS). MSP, however, is replete with problems. The 2016 Evaluation Report on Minimum Support Prices released by NITI Aayog underlined that the lack of procurement centers, closed storage facilities and delay in payments were some of the shortcomings of MSP. Lack of knowledge about MSP also contributed to farmers not being able to plan crop growing pattern ahead of sowing season and reap additional benefits from it. According to the report, despite its shortcomings farmers find MSP to be very useful and want it to continue as it provides a floor price for their produce and protects them against price fluctuations.

Despite having the capacity to pay, an individual may not have the intention or discipline to pay back the loans on a timely basis which is costly for banks. This is the behavioral aspect of credit risk and is reflected in his/her credit history. Recent delinquency, on-time payment history, leverage, default and non-default credit accounts are some metrics which give us insights into the behavior of the customer through which we can evaluate whether he/she has the intention and required discipline to pay back the loans.

In agricultural loan market, loan waivers announced by government severely impact the behavior of the agriculture households and create a moral hazard problem where farmers default on loans in expectation of loan waivers in the future. Such loan waivers undoubtedly relieve distressed farmers of their credit burden, but it negatively impacts the credit culture. Such political risks associated with the agriculture sector make banks reluctant to lend to this sector. Post the 2008 comprehensive loan waiver scheme, a survey showed that one out of every 4 respondents wanted to wait for another loan waiver (Maurer 2014).

Instead of giving out generalized loan waivers to farmers, the need of the hour is to focus on creating robust mechanism to assess credit risk in the agriculture sector which can help banks increase their reach and help bring the farmers into the formal sector. In this paper, we highlight an approach that can make this possible. We show how using farm and household characteristics we can risk rank the agriculture households by assessing their "capacity to pay." Considering the difficulties faced by farmers and banks, our model would help in bridging the gap between them. By reducing the risk associated with farm lending, it would create a potentially profitable market for banks and would make cheaper credit available to the farmers along with reducing their dependency on moneylenders.

2 Literature Review

The economic survey of 2017–18 reveals that India's agricultural sector which employs more than 50% of the population contributes only 17–18% in its total output (Economic Division 2018). Therefore, enhancement of farm mechanization is important to mitigate hidden unemployment in the sector and free up useful labor. Agricultural credit plays a pivotal role in achieving technical innovation, and therefore measures need to be taken to expand the reach of low-cost formal credit to all farmers. Abhiman Das (2009) show that direct agricultural lending has a positive and significant impact on agricultural output whereas indirect credit has an affect after a lag of one year. Therefore, despite its shortcomings like less penetration to small and marginal farmers, and paucity of medium- to long-term lending, agricultural credit plays a critical role in supporting agricultural production and hence farm incomes and livelihood. In order to lend efficiently and minimize defaults on loans, it is imperative to have a sound analytical system in place to assess creditworthiness of borrowers. There have been several studies on credit scoring models for agricultural lending which use bank or credit history data as well as farm's and borrower's characteristics to assess debt repaying capacity of farmers. Identifying low-risk customers using credit risk assessment models is important not only for reducing cost for banks but also to increase the penetration of credit to small and marginal farmers who would have otherwise been left out due to misclassification as bad customers. Bandyopadhyay (2007), using sample data of a public sector bank, developed a credit risk model for agricultural loan portfolio of the bank. With the help of bank's credit history and borrower's loan characteristics such as loan to value, interest cost on the loan, value of land and crops grown, he arrives at a logistic regression model that predicts the probability of default-defined as per the then NPA norm of the RBI, i.e., if the interest and/or installment of principals remains overdue for two harvest seasons but for a period not exceeding two and half years in the case of an advance granted for agricultural purpose. However, low sample size of the study serves as a major limitation of the model as it renders the model vulnerable to sample biases. Seda Durguner (2006), in their paper, showed that net worth does not play a significant role in predicting probability of default for livestock farms while it does matter significantly for crop farms. They develop separate model for crop and livestock farms in order to prevent misclassification errors that could arise by not differentiating between the farm types. Durguner (2007) showed using a panel data of 264 unique Illinois farmers for a five-year period, 2000–2004, that both debt-to-asset ratio and soil productivity are highly correlated with coverage ratio (cash inflow/cash outflow). Using a binomial logit regression model on 756 agricultural loan applications of French banks, Amelie Jouault (2006) show that leverage, profitability and liquidity at loan origination are good indicators of probability of default.

The studies mentioned above suffer from some severe limitations which need to be addressed for obtaining a robust credit risk model:

(1) No differentiation on geographical location and farm type: The ability of a farmer to repay depends on the income that he generates which is highly dependent on where he lives, rainfall pattern in that location, the soil type, crop grown, etc. Therefore, considering such agro-climatic factors is necessary in the model building process.
(2) Limited data sources: Bank's data would not be helpful for assessing risk of the farmers who are new to formal credit or if banks expand their direct lending to agriculture to new locations. Alternative methods to score farmers for their riskiness need to be identified as opposed to relying just on their past performance.
(3) Narrow focus of study: Credit risk from farm sector, as mentioned in the above section, can result from inability to pay that can be influenced by price risk and market risk or it could be due to indiscipline and fraudulent behavior which could result from political risk. Focusing only on the behavior of farmers on their credit account will not take into account a complete picture of the situation of the farmers, and most importantly, it would leave out those who are new to credit.
(4) Small sample size: Given the nature of diversity in India's agricultural landscape, a single bank's data cannot capture all the dimensionalities of the sector and a small sample size can lead to sample biases and cannot be applied universally.

3 Agriculture Credit in India: Trends and Current Scenario

Current mandate for Priority Sector Lending (PSL) by RBI requires all scheduled commercial banks and foreign banks to lend 18% of their Adjusted Net Bank Credit (ANBC) or Off Balance Sheet Exposure, whichever is higher, to the agriculture sector. Out of this, a sub-target requires them to lend 8% to the small and marginal farmers. As per the RBI guidelines, a small farmer is one who holds less than or equal to 1 ha of land whereas any farmer with more than 1 ha but up to 2.5 ha of land is considered to be a marginal farmer. These guidelines hold for all Scheduled Commercial Banks (SCBs) including foreign banks (RBI 2016).

Additional measures taken by the government to improve the farm credit situation include Kisan Credit Card (KCC) and Agricultural Debt Waiver and Debt Relief Scheme though their effectiveness can be debated and most of the experts consider them to be an unnecessary fiscal burden.

Despite taking the policy measures mentioned above, year-on-year growth of farm loans has gone down in past few years. After seeing a close to 40% growth rate in 2014, increase in farm credit went down to below 10% which is lowest since 2012 (Trends and Progress Report, RBI 2017–18).

As per the All India Rural Financial Inclusion Survey (NAFIS), 2016–17, by NABARD, 52.5% agricultural households had an outstanding debt at the time of the survey and out of these almost 40% households still went to non-institutional sources for their credit needs. Similar results are shown by the Situation Assessment Survey (SAS), 2013, by NSSO which shows a dependence of 44% households on non-institutional sources (please refer to Table 1). Even though two surveys have different samples, this indicates that the share of non-institutional sources has remained almost same from 2013 to 2016–17 and additionally corroborates the fact that growth in institutional credit has remained stagnant.

With flexible lending terms and often no collateral required, agricultural households continue to borrow from informal sources (moneylenders, friends and family).

Despite the exorbitant interest rates, which can go as high as 4 times the interest rates charged by the formal sources (refer to Table 2), moneylenders continue to cater to the credit needs of close to 11% of the farm borrowers (NAFIS 2016–17). This, including the reasons mentioned above, could be due to various factors including the availability of credit for personal reasons such as marriage. Another reason for this could be the unavailability of formal sources of credit. As per SAS 2013, for the agricultural households which owned less than 0.01 ha of land, only about 15% of

Table 1 Distribution of rural credit across institutional and non-institutional sources (in %)

Type of credit	1951	1961	1971	1981	1991	2002	2012
Institutional credit	7.2	14.8	29.2	61.2	64	57.1	56
Non-institutional credit	92.8	85.2	70.8	38.8	36	42.9	44

Table 2 Distribution of outstanding cash debt as per the rate of interest charged

Rate of interest (%)	Distribution of outstanding cash debt			
	Rural		Urban	
	Institutional	Non-institutional	Institutional	Non-institutional
Nil	0.8	18.3	0.4	27
<6	7.1	2.3	1.5	1.1
10-June	26	0.4	14.5	0.9
12-October	12.9	0.7	41.6	1.2
15-December	42.6	4.1	34.1	7.7
15–20	7.3	5.6	6.2	4.3
20–25	**2.1**	**33.9**	**1.2**	**27.3**
25–30	**0.1**	**0.6**	**0.2**	**0.3**
>30	**1**	**34.1**	**0.4**	**30.2**

loans were sourced from institutional lenders. On the other hand, this number was as high as 79% for farmers with more than 10 ha of land.

Most of this farm lending continues to be done by public sector banks. As of December 2016, private sector lent out 9.5% of the total loans whereas public sector lent out 85% of the total loans (Credit Bureau Database 2018). Private players, including foreign banks, have been reluctant to lend to farmers. For the year 2017–18, private and foreign banks met their PSL targets but did not meet their sub-targets of 8% lending to the small and marginal farmers (Trends and Progress Report, RBI 2017–18).

One major reason for this reluctance is the rise in bad loans coming from this sector. Between 2012 and 2017, bad loans in agriculture sector have jumped by 142.74% (Financial Express Online 2018). One reason behind this jump is the farm loan waiver announced by the central government. Subvention schemes, a subsidy provided by the government on interest rate, are another reason why private banks find PSL challenging. Banks are mandated to charge 7% interest on loans up to 3 lakhs. A further 3% subvention is provided in case of timely payments. So effectively these loans become available to farmers at 4% interest rate (PIB, DSM/SBS/KA, release ID 169414). This scheme has recently been made available to the private sector banks since 2013–14, prior to which it was only available to public sector banks.

Another reason for the meager farm lending by the private and foreign banks is the lack of understanding of the agriculture sector as a whole. This also leads to the inability to effectively assess risk in this sector. Without a proper understanding of the sector and the understanding of risk, operating in rural and semi-urban areas can be very expensive for banks. Entering a new market requires opening of new branches, launching market-specific products and huge operating costs. Due to all these reasons, these banks stay away from the agriculture sector or do marginal amount of lending in urban areas.

4 Research Methodology

Given the challenges faced by banks in lending to the agriculture sector, we propose in this paper a holistic approach to assess credit risk of farmers using alternative data and advanced analytical techniques. The focus of this study is the farmers who are not a part of the formal credit and still rely on non-institutional sources. These farmers would not have a credit footprint available to assess their riskiness, and therefore we focus on "capacity to pay" of the farmers rather than their default behavior on their credit accounts. In this study, we have used NSSO 70th round data—Key Indicators of Situation of Agricultural Households in India to identify complete characteristics of farmers in India. This is a comprehensive dataset of agricultural households in India which are defined as households receiving at least Rs. 3000 of value from agricultural activities (e.g., cultivation of field crops, horticultural crops, fodder crops, plantation, animal husbandry, poultry, fishery, piggery, beekeeping, vermiculture, sericulture, etc.) during last 365 days, and it encompasses all the factors that reflect the then situation of farmers. The survey was conducted in two visits. Visit 1 comprises data collected in the period January 2013 to July 2013 with information collected with reference to period July 2012 to December 2012, and Visit-2 comprises data collected between August 2013 and December 2013 with information with reference to period January 2013 to July 2013. This way it covers both kharif and rabi cropping seasons. However, for our modeling purpose we use only Visit-1 data as the information on outstanding loans is captured only in the Visit-1 Survey. The NSSO data captures variables such as the kind of dwelling unit of farmers, status of ownership of land, primary and subsidiary activity of farmers, whether the household has MGNREG job cards, no. of dependents and their employment status, the kind of crop grown on the farm, size of land under irrigation, the value of sale of crop, the agency the crops are sold through (dealers, mandi, cooperative agency and government), details of expenses in inputs and whether the farmer avails MSP or not. Such a detailed dataset of farmer characteristics is very helpful in assessing whether the farmer will be able to "afford" the loan or not. Following Seda Durguner (2006), we used debt to income as a proxy to judge creditworthiness. The mean debt to income in the population is 14, while the median is 1.5. Below table shows the distribution of debt to income in the data (Refer to Table 3).

We use debt-to-income ratio of 4 as the threshold; i.e., farmers whose ratio of outstanding debt is more than 4 times the income of one cropping season are classified as bad (farmers who would default), and logistic regression technique is used to predict the probability of default for these farmers. The overall bad rate of the population with the given threshold is 26%.

We build two models in our analysis. First, we use the variables that are captured by banks (Model 1) in their agricultural loan application form. For this purpose, we use standardized loan application form for agricultural credit devised by Indian Bank's Association (IBA). This form contains the required details that need to be collected from agri-credit loan applicants. This helps banks and customers maintain uniformity in the loan applications for agricultural needs. The second model (Model

Table 3 Distribution of debt-to-income ratio in our data

Quantile	Estimate
100% max	21,300
0.99	186.96
0.95	51.07
0.9	18.18
75% Q3	4.8
50% median	1.55
25% Q1	0.53
0.1	0.19
0.05	0.1
0.01	0.03
0% min	0

2) that we built considered both, the details already captured by the bank along with the additional features created from the NSS 70th round data. We use information value (IV), which tells how well my variable is able to distinguish between good and bad customers, to select important or predictive variables in the model. The variables whose IV was between 0.02 and 0.5 were then binned using weight of evidence (WOE). Variables with similar WOE were combined in a bin because they have similar distribution of events and nonevents. In this way, we transformed continuous independent variable to a set of groups/bins. We then built a logistic regression model to obtain probability of default using WOE of independent variables.

We find that Model 2 performs better than Model 1 in terms of Gini, KS and rank ordering. The results of the model are discussed in the next section.

5 Results

Our model gives a comprehensive set of variables which includes farmer's demographic features, agro-climatic factors and cropping patterns that describe his/her ability to pay. Variables like highest value crop grown and whether the farmer faced crop loss during the last one year capture the farming pattern for the farmer and explain how the recent trend of farming has been for the farmer. Whether the farmer has taken technical advice or not shows if farmer has access and willingness to incorporate new techniques in his farming. Our model covers both the endowment and behavior-related variables of the farmers.

The following tables give the resultant significant variables in both the models:

1. Model 1:

Analysis of maximum likelihood estimate		
Parameter	Sign of coefficient	Pr > ChiSq
Intercept	Positive	<0.0001
Primary income source	Negative	<0.0001
Percentage of land cultivated	Negative	<0.0001
Percentage of expense on machine hiring	Negative	<0.0001
Percentage of expense on fertilizers and chemicals	Negative	<0.0001
Percentage of expense on seeds	Negative	<0.0001
Number of male members in the family	Negative	<0.0001
Count of members between the age of 18 and 60 years	Negative	<0.0001
Age of the household head	Negative	<0.0001

2. Model 2:

Analysis of maximum likelihood estimate		
Parameter	Sign of coefficient	Pr > ChiSq
Intercept	Negative	<0.0001
Whether technical advice taken or not	Negative	<0.0001
Whether farmer suffered crop loss in the last season or not	Negative	<0.0001
Primary income source	Negative	<0.0001
Segment of the highest value crop grown by the farmer	Negative	<0.0001
Rainfall as a percentage of average rainfall in the district	Negative	<0.0001
Percentage of cultivated land	Negative	<0.0001
Percentage of expense on machine hiring	Negative	<0.0001
Percentage of expense on seeds	Negative	<0.0001
Number of male members in the family	Negative	<0.0001
Age of the household head	Negative	<0.0001

Capturing these additional variables in our model for assessing the risk of farmers gives us a lift of almost 7% in the Gini coefficient (from 41.7 to 48%); i.e., it improves the model accuracy from 70 to 75%. Bad rate distribution for our model goes from 58.87% in the lowest decile (highest risk decile) to 5.48% in the highest decile (lowest risk decile). On the other hand, using the variables already captured by banks, bad rate ranged from 55.27 to 8.02%. The below graph shows the risk ranking across deciles for both the models. We observe a break in the risk ranking of Model 1 at decile 7, whereas Model 2 holds perfectly across all the deciles. Refer to Appendix for detailed tables and to Appendix 2 for a note on Gini and KS Summary Statistic.

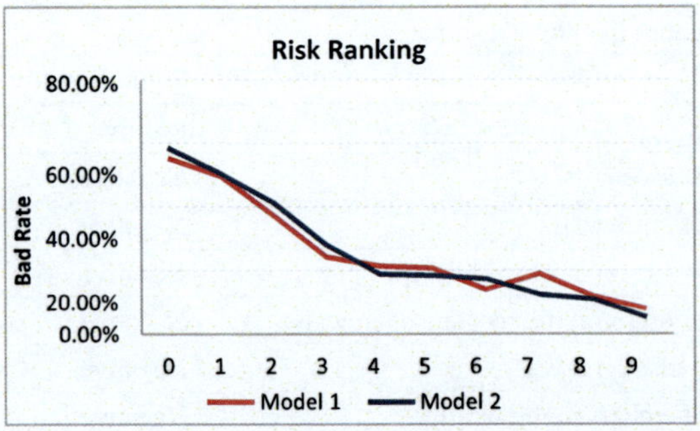

Model 2 provides a significant decrease in risk levels in comparison with Model 1 assuming that banks keep their approval rates constant across models. For example, if a bank decides to approve 19% of the credit applications received, it would face a 20% lower risk of default using Model 2 as compared to Model 1. This would allow banks to curb their bad rates and would be welfare generating for both the banks and the farmers.

3. Out-of-sample validation results: To assess the stability of our models across samples, we validated them on a randomly selected 30% sample of the development data. The result is given below:

Samples	Gini (%)	KS (%)
Model 1	41.51	32.76
Model 2	47.68	37.48

4. To check the applicability of Model 2 on different farmer segments based on their land holding, we validated the model on marginal, small and other farmers as defined by RBI. The model holds well in these segments in terms of rank ordering, Gini and KS, but some variables do not rank order in "Other Farmers" segment. The result is given below:

Samples	Gini (%)	KS (%)
Small farmers	47.15	36.27
Marginal farmers	47.24	37.30
Others	51.62	40.13

6 Policy Implications

A policy aspect that comes out from this analysis is that this model would allow the government to figure out the population they need to focus on for their policy measures. Farmers who get identified with a lower capability to pay using Model 2 become the target population for the government policies. Also, the model variables on which they did not do well define the areas where government needs to focus to bring those farmers to the formal financial sector. For example, if a large number of farmers in a district are identified to have a lower capability to repay due to having not taken technical advice or because of having suffered crop loss in the last farming season, this defines the focus area for the government to work upon. Here, they need to improve the availability of technical advice to the farmers and work on reasons of crop loss. Hence, above model would serve the dual purpose of helping both the banks and the government. Even though banks need to keep lending at the reduced interest rates as per the government policies, using this model they can identify the population with a lower risk of default. At the same time, government can form specific policies based on the needs of the farmers and help bring them to the formal credit market.

7 Limitations and Conclusion

Even though our model brings out results which can help both the banking sector and government, we do not claim that our model is free of any limitations or has no scope for improvement. Considering the type of data that has been used for this model, there is an inherent risk of endogeneity to occur in the analysis and it needs to be accounted for in the model building process. Also, the variables in Model 2 are not easily verifiable and it would require banks to invest in proper due diligence of their agricultural loan applicants.

On the basis of above information, it is understandable that farming sector needs special attention when it comes to credit facilities. Existing schemes and facilities have been unable to fulfill the credit needs of this sector. Generalized loan waivers announced time, again have put a financial burden on the economy and are not a solution in the long run. Our model shows that if banks capture specific information about farmer characteristics and consider agro-climatic conditions like rainfall in their lending decisions, they can reduce the delinquencies from this sector. In this way, agricultural lending can be made much more efficient and the level of financial inclusion of farmers can be improved.

Appendix 1

Results and approval strategy for Model 1:

Decile	Good	Bad	Total	Bad rate (%)	% population
0	16,971	20,974	37,945	55.27	9.48
1	20,799	20,403	41,202	49.52	10.30
2	26,315	15,600	41,915	37.22	10.47
3	25,543	8251	33,794	24.42	8.45
4	42,643	11,776	54,419	21.64	13.60
5	23,761	6272	30,033	20.88	7.51
6	32,621	5360	37,981	14.11	9.49
7	36,168	8574	44,742	19.16	11.18
8	32,855	4569	37,424	12.21	9.35
9	37,436	3262	40,698	8.02	10.17
			Overall bad rate		**26.25**
				Gini	**41.4**

Approval Strategy:

Decile	Good	Bad	% population	Default rate (%)	Approved population (%)
0	16,971	20,974	9.48	26	100.00
1	20,799	20,403	10.30	23	90.52
2	26,315	15,600	10.47	20	80.22
3	25,543	8251	8.45	17	69.75
4	42,643	11,776	13.60	16	61.30
5	23,761	6272	7.51	15	47.70
6	32,621	5360	9.49	14	40.20
7	36,168	8574	11.18	13	30.70
8	32,855	4569	9.35	10	19.52
9	37,436	3262	10.17	8	10.17

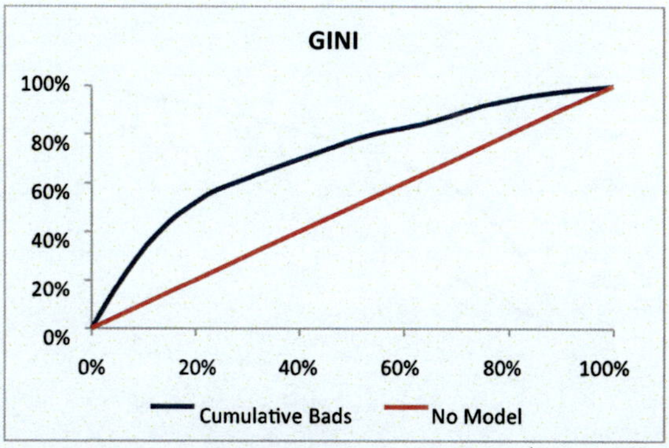

Results and approval strategy for Model 2:

Decile	Good	Bad	Total	Bad rate (%)	% population
0	16,596	23,349	39,945	58.45	9.98
1	20,542	20,808	41,350	50.32	10.33
2	21,034	14,688	35,722	41.12	8.93
3	34,241	13,398	47,639	28.12	11.91
4	28,493	6698	35,191	19.03	8.79
5	30,892	7029	37,921	18.54	9.48
6	35,153	7484	42,637	17.55	10.66
7	36,267	5212	41,479	12.57	10.37
8	34,434	4302	38,736	11.11	9.68
9	37,460	2073	39,533	5.24	9.88
			Overall bad rate		**26.25**
				Gini	**47.9**

Approval Strategy:

Decile	Good	Bad	% population	Default rate (%)	Approved population (%)
0	16,596	23,349	9.98	26	100.00
1	20,542	20,808	10.33	23	90.02
2	21,034	14,688	8.93	19	79.68
3	34,241	13,398	11.91	16	70.76
4	28,493	6698	8.79	14	58.85
5	30,892	7029	9.48	13	50.06
6	35,153	7484	10.66	12	40.58
7	36,267	5212	10.37	10	29.93
8	34,434	4302	9.68	8	19.56
9	37,460	2073	9.88	5	9.68

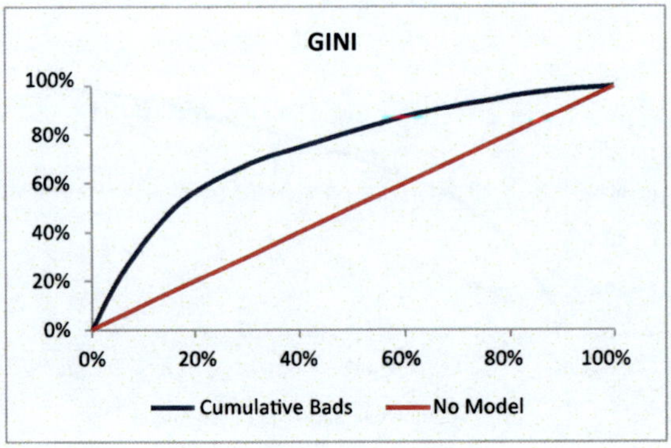

Appendix 2: Key Summary Statistic

Kolmogorov–Smirnov Test

The test was based on the following hypothesis:

$H0$: The Validation and Development Samples Have the Same Distribution.

$H1$: The Validation and Development Samples Do not Have the Same Distribution.

$D_{\max}$ is the maximum absolute difference between the score distributions for the two unweighted samples.

Applying the Kolmogorov–Smirnov test, we accept H0 if

$$D_{\max} < D_{\text{crit}} \quad \text{where } D_{\text{crit}} = 100 * K\sqrt{\frac{1}{M} + \frac{1}{N}}$$

where

M is the total number in the development sample.

N is the total number in the validation sample.

K is the Kolmogorov–Smirnov statistic.

Gini Coefficient

The Gini coefficient is a measure of the power of a scorecard. The higher the Gini, the stronger the scorecard. A scorecard with no discrimination would have a Gini of zero; a perfect scorecard would have a Gini of 100%. A Gini is calculated by comparing the cumulative number of goods and bads by score. Graphically, it is the area between the two lines on the curve below (XYZ) expressed as a percentage of the maximum possible (*XYW*). The two axes on the graph are cumulative percentage of goods (*y*-axis) and cumulative percentage of bads (*x*-axis). Graphical representation of the Gini coefficient:

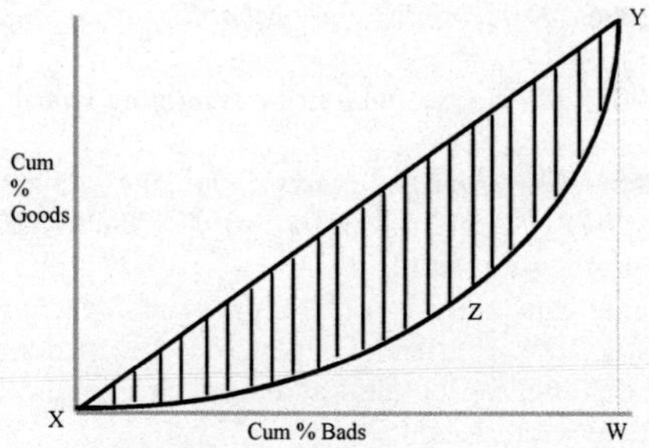

The Gini coefficient is calculated as follows:

Given that:

b_i Cumulative percentage of bads at a given score

g_i Cumulative percentage of goods at a given score

S_n *N*th score in the score distribution.

The area under the curve (the unshaded area, not enclosed within *Z*) for a given score is defined as:

$$A_i = \frac{1}{2}(b_i - b_{i-1}) * (g_i + g_{i-1})$$

The total area not defined by the curve is equal to:

$$A_g = \sum_{i=S_1}^{S_n} A_i$$

And the area defined by the triangle *XYW* is equal to:

$$A_{X\hat{Y}W} = \frac{1}{2}(100 * 100) = 5000$$

The Gini coefficient is then calculated as:

$$g = \frac{\left(A_{X\hat{Y}W} - A_g\right)}{A_{X\hat{Y}W}}$$

References

Abhiman Das, M. S. (2009). *Impact of agricultural credit on agricultural production: An empirical analysis in India.* Reserve Bank of India Occasional Papers.

Amelie Jouault, A. M. (2006). *Determining the probability of default of agricultural loans in a French bank.*

Bandyopadhyay, D. A. (2007). *Credit risk models for managing bank's agricultural loan portfolio.* MPRA.

Dinesh, U. (2013, March 17). *Private banks reluctant about rural lending.* Retrieved from Livemint: https://www.livemint.com/Industry/TWPW6KvEmOksgvzfLElCdO/Private-banks-reluctant-about-rural-lending.html.

Durguner, S. (2007). A panel data analysis of the repayment capacity of farmers. In *2007 Annual Meeting*, July 29-August 1, 2007, Portland, Oregon, No 9361, American Agricultural Economics Association (New Name 2008: Agricultural and Applied Economics Association). https://EconPapers.repec.org/RePEc:ags:aaea07:9361.

Economic Division, D. o. (2018). Economic Survey 2017–18 Volume 1. Ministry of Finance, Government of India.

Experian India Pvt Ltd. (2018). Credit Bureau Database. Experian India Pvt Ltd.

Financial Express Online. (2018, January 10). Retrieved from Financial Express: https://www.financialexpress.com/industry/banking-finance/rbi-data-agriculture-bad-loans-jump-by-23-per cent-thanks-to-farmers-loan-waiver-demonetisation/1008842/.

Indiastat. (2008, 2017). Banks/sector wise NPAs of private sector banks in India. Indiastat.

Iyer, A. (2019, January 02). *Farm loan defaults rise as banks brace for big write-offs.* Retrieved from Livemint: www.livemint.com.

Jayakumar, T. (2018, May 03). *Time to do away with priority lending norms.* Retrieved from liveMint: https://www.livemint.com/Opinion/QwGGfDgozR6COUYFyW3PTP/Time-to-do-away-with-priority-lending-norms.html.

Lahiri, A., & Mookherjee, D. (2015, December 14). *Transforming Indian agriculture: The role of credit policy.* Retrieved from Ideas for India: https://www.ideasforindia.in/topics/agriculture/transforming-indian-agriculture-the-role-of-credit-policy.html.

Maurer, K. (2014). Where is the risk? Is agricultural banking really more difficult than other sectors? In *Finance for food: Towards new agricultural and rural finance.*

NAFIS. (2016–17). All India Rural Financial Inclusion Survey (NAFIS) 2016–17, NABARD.

Niti Aayog. (2016). *Evaluation report on efficacy of minimum support prices on farmers.* New Delhi: Development Monitoring and Evaluation Office.

Peter Binswanger, H., & Khandkar, R. S. (1992, October). The impact of formal finance on the rural economy of India. *Journal of Development Studies.*

Pradhan, N. C. (2013, May). Persistence of informal credit in Rurak India: Evidence from 'All India Debt and investment survey'. *RBI WPS (DEPR).*

Press Information Bureau (PIB), DSM/SBS/KA, release ID 169414.

PTI. (2014, August 26). *Nabard's rural infrastructure development fund needs a relook, says RBI official.* Retrieved from Economictimes: https://economictimes.indiatimes.com/news/economy/policy/nabards-rural-infrastructure-development-fund-needs-a-relook-says-rbi-official/articleshow/40928706.cms.

RBI, F. S. (2016). *Financial stability report.*

Sayantan, B. (2015, March 19). *Surge in credit not benefiting small farmers.* Retrieved from Livemint: https://www.livemint.com/Politics/xbk7sf9N4gy0jjStXyLoiJ/Surge-in-credit-not-benefiting-small-farmers.html.

Seda Durguner, P. J. (2006, July). *Credit scoring models: A comparison between crop and livestock farms.* Long Beach, California, United States of America.

Standardized common loan application form for agricultural credit, Indian bank's Association (IBA), No. SB/Cir/AGRI/480

Trends and Progress Report, RBI. (2017–18). *Trends and progress report.* RBI.

Application of Association Rule Mining in a Clothing Retail Store

Akshay Jain, Shrey Jain, and Nitin Merh

1 Introduction

Retailing is a sale of goods or commodities in small quantities directly to the customers. Retailers provide important functions that increase the value of the products and services they sell to customers. These value-creating functions are providing assortment of products and services, breaking bulk, holding inventory and providing services and experiences (Avcilar et al. 2014). Retail industry is a business which has a high market concentration but has the potential to generate high profits if managed properly with the help of certain tools. The tools which contribute to the success of retail store help in understanding the consumer behavior, buying pattern, relativity between the products. Consumer behavior is a consumer activity in deciding to purchase, use, as well as consume the purchased goods and services including the customer factors which can give a rise to their decisions whether to purchase and use products (Kurniawan et al. 2018).

Data mining is the process of automatically discovering useful information in large data repositories. Data mining techniques are deployed to scour large databases in order to find useful patterns that might otherwise remain unknown (Tan et al. 2016). Association rule mining, first introduced by Agrawal et al. (1993), is useful for discovering interesting relationships hidden in large datasets. The uncovered relationship can be represented in the form of association rule or sets of frequent items (Tan et al. 2016). Market basket analysis is a technique that analyzes customer

A. Jain · S. Jain
SVKM's Narsee Monjee Institute of Management Studies (NMIMS), Indore, India
e-mail: akshay2602jain@gmail.com

S. Jain
e-mail: shrey.1425@gmail.com

N. Merh (✉)
Jaipuria Institute of Management Indore, Indore, India
e-mail: nitinmerh0812@gmail.com

A. K. Laha (ed.), *Applied Advanced Analytics*, Springer Proceedings in Business and Economics, https://doi.org/10.1007/978-981-33-6656-5_9

buying habits with the help of associations between the products that the customer place in their shopping basket. This will help the retailers to develop marketing strategies by gaining insights into the items that are frequently purchased with each other.

Business analytics is a scientific process of transforming data into insight for making better decisions. Business analytics is used for data-driven or fact-based decision making which is often seen as more objective than other alternatives. It is of three types: descriptive, predictive and prescriptive; out of these, this research is focusing on predictive analytics (Cochran et al. 2015). Predictive analytics includes variety of statistical techniques from modeling, data mining and machine learning that analyze current and historical data to make predictions about future outcomes. Prediction helps organization in making right decision at right time by right person as there is always time lag between planning and actual implementation of the event.

"Try Us," a new retail clothing store located in Indore, Madhya Pradesh, is selected as the store under study. It sells multiple brands for men. The organization wants to expand its business and is planning to open another store on a larger scale.

Apriori algorithm is used for mining datasets for association rules. The name Apriori is used because it uses prior knowledge of frequent item properties. Apriori algorithm uses bottom-up approach where frequent subsets are extended one by one. In Apriori algorithm, iterative approach is used where k frequent item sets are used to find $k + 1$ item sets. It is generally used in market basket analysis as it is useful in finding the relationship between two products. Apriori makes some assumptions like:

- All subsets of a frequent item set must be frequent.
- If an item set is infrequent, all its supersets will be infrequent.

While applying Apriori algorithm, the standard measures are used to assess association rules. These rules are the support and confidence value. Both are computed from the support of certain item sets. For association rules like $A \rightarrow B$, two criteria are jointly used for rule evaluation. The support is the percentage of transactions that contain $A \bigcup B$ (Agrawal et al. 1993; Avcilar et al. 2014). It takes the form support $(A \rightarrow B) = P(A \bigcup B)$. The confidence is the ratio of percentage of transactions that contain $(A \bigcup B)$ to the percentage of transactions that contain A. It takes the form confidence $(A \rightarrow B) = P(B \backslash A) =$ support$(A \bigcup B)$/support (A). Rules that satisfy both a minimum support threshold (min_sup) and minimum confidence threshold (min_conf) are called strong (Avcilar et al. 2014).

Primary objective of the study is to understand the buying pattern of the customer and to study and analyze proper basket (combos) of products for cross-selling and upselling. Another objective is to explore the relativity between the products for applying the optimal design layout for the clothing retail store.

1.1 Literature Review

Research work done by Kurniawal et al. (2018) suggests that market basket analysis performs better results over association rule mining using Apriori algorithm. The research done by Tatiana et al. (2018) on a study of integrating heterogeneous data sources from a grocery store based on market basket analysis, for improving the quality of grocery supermarkets, shows positive results for increasing the performance of the store.

Szymkowiak et al. (2018) propose theoretical aspects of market basket analysis with an illustrative application based on data from the national census of population and housing with respect to marital status, through which it was made possible to identify relationships between legal marital status and actual marital status taking into account other basic socio-demographic variables available in large datasets. Study (Roodpishi et al. 2015) conducted on various demographic variables for an insurance company in the city of Anzali, Iran, provides various associations with clients of an insurance company. The study used association rules and practice of insurance policy to find hidden patterns in the insurance industry.

In the study done by Sagin et al. (2018), market basket analysis was conducted on a data of a large hardware company operating in the retail sector. Both the Apriori and FP growth algorithms (Sagin et al. 2018) were run separately and their usefulness in such a set of data was compared. When both the algorithms were compared in terms of performance, it was seen that FP growth algorithm yielded 781 times faster results but resulting rules showed that FP growth algorithm failed to find the first 14 rules with high confidence value. In the study done by Srinivasa Kumar et al. (2018), product positioning of a retail supermarket based at Trichy, Tamil Nadu, was examined using data mining to identify the items sets that were bought frequently and association rules were generated. The study done by Santarcangelo et al. (2018) focused on visual market basket analysis with the goal to infer the behavior of the customers of a store with the help of dataset of egocentric videos collected during customer's real shopping sessions. They proposed a multimodal method which exploited visual, motion and audio descriptors and concluded that the multimodal method achieved an accuracy of more than 87%. In the study done by Avcilar et al. (2014), association rules were estimated using market basket analysis and taking support, confidence and lift measures. These rules helped in understanding the purchase behavior of the customers from their visit to a store while purchasing similar and different product categories. The objective of the research study done by Seruni et al. (2005) was to identify the associated product, which then were grouped in mix merchandise with the help of market basket analysis. The association between the products was then used in the design layout of the product in the supermarket.

1.2 Pricing Intelligence

Pricing intelligence consists of tracking, monitoring and analyzing pricing data to understand the market and make educated pricing changes at speed and scale (Ballard 2018). Pricing intelligence can help in determining the effective price for various products that will give an edge over the competitors and also help in boosting up the sales. If used smartly, it can also act as a tool to clear the pending stock in an outlet. Pricing of a product can be determined by keeping various factors in mind such as time duration of a particular product in a shelf and competitor's price for the same product. Discount percentage could also be determined by the time a product is on the shelf.

2 Methodology

In the current study, an attempt is made to find the relationship between the different products using Apriori algorithm.

At the first stage, data is preprocessed and transformed, values are handled and the data is cleaned before selecting the components. After transforming the data, Apriori algorithm was used to find the relationship between different apparels using association rules. The data is analyzed on the basis of results obtained from Frontline Analytic Solver® Data Mining (XLMiner).

Various parameters used for evaluation of the model are antecedent support (if part) which is the number of transactions in which item/s is present, consequent support (then part) which is number transactions in which item/s is present, support which is number of transactions that include all items in the antecedent and consequent. Antecedent (the "if" part) and the consequent (the "then" part), an association rule contains two numbers that express the degree of uncertainty about the rule.

The first number is called the support which is simply the number of transactions that include all items in the antecedent and consequent. The second number is confidence which is the ratio of the number of transactions that include all items in the consequent as well as the antecedent (namely, the support) to the number of transactions that include all items in the antecedent.

Lift is another important parameter of interest in the association analysis. It is the ratio of confidence to expected confidence. A lift ratio larger than 1.0 implies that the relationship between the antecedent and the consequent is more significant. Larger the lift ratio, the more significant the association. The following are the parameters used to evaluate the model:

- Support—support which is simply the number of transactions that include all items in the antecedent and consequent.
- Confidence = (no. of transactions with antecedents and consequent item sets)/(no. of transactions with antecedents item sets).

- Benchmark confidence = (number of transactions in consequent item sets)/(number of transactions in database).
- Lift ratio = confidence/benchmark confidence.

In Frontline Analytic Solver® Data Mining (XLMiner), minimum support transaction and confidence percentage controlled parameters were used for designing the model and checking the performance of the data mining.

3 Data

The data used for this paper is collected from "Try Us" a multi-brand retail outlet in Indore, Madhya Pradesh, for a period during November 26, 2017, to September 19, 2018. The collected data is used for the study of association between different products, and inferences generated can then be used to arrange shelves in a better way when planned for a bigger retail store. The data collected includes bill number, date, brand name, size, amount, GST, item type which was refined according to our purpose to bill number, brand name and item type.

For applying Apriori algorithm on binary data format, the data was first converted to binary format where if a product was purchased it was recorded as 1 and if no purchase was made then it was recorded as 0. In total, there are 29 columns and 13,065 rows which were refined to 29 columns and 6008 rows since multiple items purchased by a customer were recorded in multiple rows. The brands that were not present in the store from November 26, 2017, were not taken into consideration. Therefore, 185 rows were deleted out of 6193 rows. Multiple purchases made by a single customer were merged in a single row.

3.1 Data Analysis, Results and Findings

Main objective of the study is to study the buying pattern of the customer and to analyze proper basket (combos) of products for cross-selling and upselling. Another objective is to study the association between the products for applying the optimal design layout for the retail store. In the paper, association rule mining through Apriori algorithm is used to find baskets of products which are purchased together. A total of 5223 transactions are included in the analysis. Using combination of various minimum support transactions and minimum confidence percentage, the following results are derived:

Case I

Association rules: fitting parameters	
Method	Apriori
Min support	50
Min confidence	20

Rules

Rule ID	Antecedent	Consequent	A-support	C-support	Support	Confidence	Lift ratio
Rule 1	[TSHIRTSA]	[SHIRTSB]	660	1994	143	21.67	0.57
Rule 2	[TSHIRTG]	[SHIRTSB]	547	1994	116	21.21	0.56
Rule 3	[TSHIRTN]	[SHIRTSN]	173	1121	59	34.10	1.59
Rule 4	[SHIRTSN]	[SHIRTSB]	1121	1994	259	23.10	0.61
Rule 5	[SHIRTSF]	[SHIRTSB]	307	1994	67	21.82	0.57
Rule 6	[SHIRTSJ]	[SHIRTSB]	150	1994	51	34.00	0.89
Rule 7	[SHIRTSA]	[SHIRTSB]	257	1994	70	27.24	0.71
Rule 8	[JEANSC]	[SHIRTSB]	623	1994	240	38.52	1.01
Rule 9	[JEANSN]	[SHIRTSB]	401	1994	87	21.70	0.57
Rule 10	[JEANSO]	[SHIRTSB]	234	1994	101	43.16	1.13
Rule 11	[JEANSM]	[SHIRTSB]	208	1994	70	33.65	0.88
Rule 12	[TROUSERD]	[SHIRTSB]	438	1994	168	38.36	1.00
Rule 13	[JEANSN]	[SHIRTSN]	401	1121	187	46.63	2.17
Rule 14	[TROUSERD]	[SHIRTSN]	438	1121	96	21.92	1.02
Rule 15	[JEANSF]	[SHIRTSF]	196	307	60	30.61	5.21

Lift ratio—Lift value of an association rule is the ratio of the confidence of the rule and the expected confidence.

Confidence percentage—The confidence of an association rule is a percentage of number of transactions with antecedents and consequent item sets divided by number of transactions with antecedents item sets.

In **case I,** the rules having lift ratio more than 1 are rule 3, rule 8, rule 10, rule 12, rule 13, rule 14, rule 15. A brief description of these rules is given below.

Rule 3

A customer who purchases T-shirt N (Mufti) purchases a shirt N (Mufti).

Rule 8

A customer who purchases jeans C (Nostrum) purchases a shirt B (Ecohawk).

Rule 10

A customer who purchases jeans O (Revit) purchases a shirt B (Ecohawk).

Rule 12

A customer who purchases trouser D (Sixth Element) purchases a shirt B (Ecohawk).

Rule 13

A customer who purchases jeans N (Mufti) purchases a shirt N (Mufti).

Rule 14

A customer who purchases trouser D (Sixth Element) purchases a shirt N (Mufti).

Rule 15

A customer who purchases jeans F (US Polo) purchases a shirt F (US Polo).

The products with the lift ratio between 1 and 2 should be clubbed and kept together on the same shelf. For example, T-shirts N (Mufti) and shirt N (Mufti), jeans C (Nostrum) and shirt B (Ecohawk), jeans O (Revit) and shirt B (Ecohawk), trouser D (Sixth Element) and shirt B (Ecohawk), trouser D (Sixth Element) and shirt N (Mufti) should be clubbed and kept together.

Similarly, the products with the lift ratio between 2 and 6 should be clubbed and kept together on the same shelf.

Case II

Association rules: fitting parameters	
Method	Apriori
Min support	100
Min confidence	20

Rules:

Rule ID	Antecedent	Consequent	A-support	C-support	Support	Confidence	Lift ratio
Rule 1	[TSHIRTSA]	[SHIRTSB]	660	1994	143	21.67	0.57
Rule 2	[TSHIRTG]	[SHIRTSB]	547	1994	116	21.21	0.56
Rule 3	[SHIRTSN]	[SHIRTSB]	1121	1994	259	23.10	0.61
Rule 4	[JEANSC]	[SHIRTSB]	623	1994	240	38.52	1.01
Rule 5	[JEANSO]	[SHIRTSB]	234	1994	101	43.16	1.13
Rule 6	[TROUSERD]	[SHIRTSB]	438	1994	168	38.36	1.00
Rule 7	[JEANSN]	[SHIRTSN]	401	1121	187	46.63	2.17

In **case II,** the rules having lift ratio more than 1 are rule 4, rule 5, rule 6, rule 7. A brief description of these rules is given below.

Rule 4

A customer who purchases jeans C (Nostrum) purchases a shirt B (Ecohawk).

Rule 5

A customer who purchases jeans O (Revit) purchases a shirt B (Ecohawk).

Rule 6

A customer who purchases trouser D (Sixth Element) purchases a shirt B (Ecohawk).

Rule 7

A customer who purchases jeans N (Mufti) purchases a shirt N (Mufti).

Thus, jeans C (Nostrum) and shirt B (Ecohawk), jeans O (Revit) and shirt B (Ecohawk), trouser D (Sixth Element) and shirt B (Ecohawk), jeans N (Mufti) and shirt N (Mufti) should be clubbed and kept together.

Case III

Association rules: fitting parameters	
Method	Apriori
Min support	150
Min confidence	20

Rules:

Rule ID	Antecedent	Consequent	A-support	C-support	Support	Confidence	Lift ratio
Rule 1	[SHIRTSN]	[SHIRTSB]	1121	1994	259	23.10	0.61
Rule 2	[JEANSC]	[SHIRTSB]	623	1994	240	38.52	1.01
Rule 3	[TROUSERD]	[SHIRTSB]	438	1994	168	38.36	1.00
Rule 4	[JEANSN]	[SHIRTSN]	401	1121	187	46.63	2.17

In **case III,** the rules having lift ratio more than 1 are rule 2, rule 3, rule 4. A brief description of these rules is given below.

Rule 2

A customer who purchases jeans C (Nostrum) purchases a shirt B (Ecohawk).

Rule 3

A customer who purchases trouser D (Sixth Element) purchases a shirt B (Ecohawk).

Rule 4

A customer who purchases jeans N (Mufti) purchases a shirt N (Mufti).

Thus, jeans C (Nostrum) and shirt B (Ecohawk), trouser D (Sixth Element) and shirt B (Ecohawk), jeans N (Mufti) and shirt N (Mufti) should be clubbed and kept together.

From the data, it was observed that shirt B (Ecohawk) and shirt N (Mufti) had a very strong association as they both were sold together for 259 times. Similarly, shirt B (Ecohawk) and jeans C (Nostrum) were sold together for 240 times.

Figure 1 gives a radar chart of what products are purchased together:

Different colors represent different types of products that were available in the store. Each concentric circle represents 50 transactions. The following suggestions can be given to the entrepreneur after analyzing the data.

- On the basis of market basket analysis, products which are sold together with high frequency like shirt B (Ecohawk) and jeans N (Mufti) should be kept near to each other such that it reduces the handling time of the customer by the salesperson. Furthermore, baskets can be developed using the analysis done above.
- Products which have a low frequency should be near to the products that are preferred more by the customer with some dynamic discount pattern so as to increase the sales of the low frequency products.

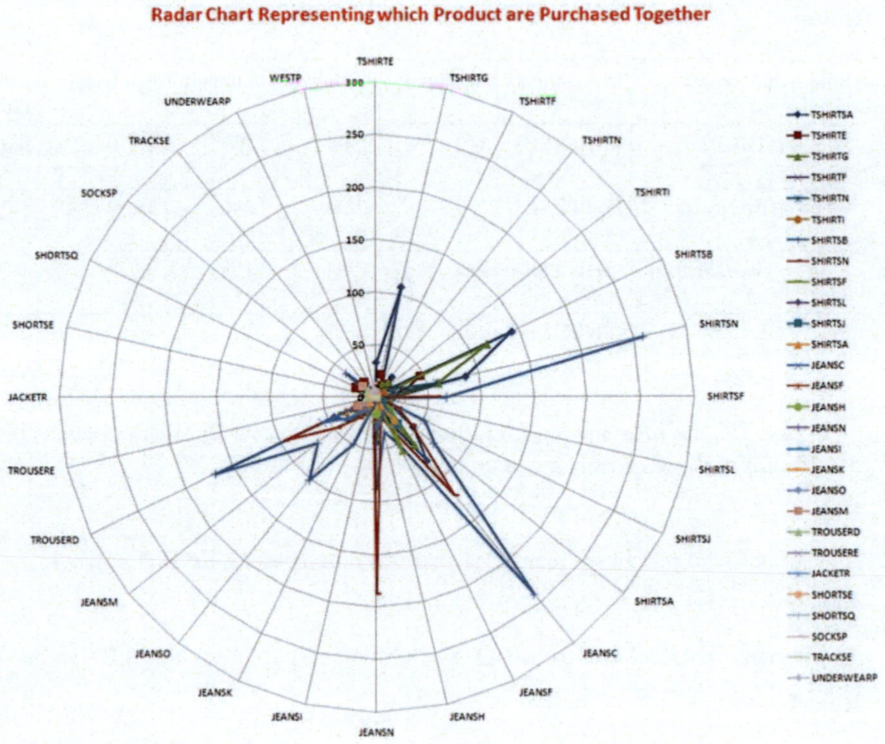

Fig. 1 Radar chart

4 Conclusion

Applying market basket analysis, some meaningful patterns were obtained which would help the entrepreneur in planning the discount pattern of the products and also making various combos according to the rules generated in order to increase the sales. It will also help in preparing the shelf design for the new store so as to minimize the product searching time by the customer.

Appendix

Details containing brand name and code used in the research.

Serial No.	Code	Brand
1	A	Cookyss
2	B	Ecohawk

(continued)

(continued)

Serial No.	Code	Brand
3	C	Nostrum
4	D	Sixth Element
5	E	Status Que
6	F	US Polo
7	G	Stride
8	H	Killer
9	I	Ed Hardy
10	J	Yankee
11	K	Vogue Raw
12	L	Delmont
13	M	Rookies
14	N	Mufti
15	O	Revit
16	P	UCB
17	Q	Beevee
18	R	Silver Surfer
19	T	Status Quo
20	U	M Square
21	W	Got It
22	X	Borgoforte
23	Y	Fort Collins
24	Z	Okane

References

Agrawal, R., Imieliński, T., & Swami, A. (1993). Mining association rules between sets of items in large databases. *Association of Computing Machinery (ACM) SIGMOD Record*, *22*(2). Newyork, USA. https://doi.org/10.1145/170036. 170072. ISSN-0163-5808.

Avcilar, M. S., & Yakut, E. (2014). Association rules in data mining: An application on a clothing and accessory specialty store. *Canadian Social Science, 10*(3), 75–83.

Ballard, A. (2018, August 06). Pricing intelligence: What it is and why it matters. Retreived from https://www.mytotalreatil.com/article/pricing-intelligence-what-it-is-and-why-it-matters/. Date of downloading January 31, 2019.

Cochran, C., Ohlmann, F., Williams, A. S. (2015). *Essentials of business analytics*, pp. 323–324. ISBN-13: 978-81-315-2765-8.

Kurniawan, F., Umayah, B., Hammad, J., Mardi, S., Nugroho, S., & Hariadi, M. (2018). Market basket analysis to identify customer behaviors by way of transaction data. *Knowledge Engineering and Data Science KEDS, 1*(1), 20–25.

Roodpishi, M. V., & Nashtaei, R. (2015). Market basket analysis in insurance industry. *Management Science Letters, 5*, 393–400.

Sagin, A. N., & Ayvaz, B. (2018). Determination of association rule with market basket analysis: An application of the retail store. *Southeast Europe Journal of Soft Computing, 7*(1), 10–19.

Santarcangelo, V., Farinella, G. M., Furnari, A., & Battiato, S. (2018). Market basket analysis from egocentric videos. *Pattern Recognition Letters, 112*, 83–90.

Srinivasa Kumar, V., Renganathan, R., VijayBanu, C., & Ramya, I. (2018). Consumer buying pattern analysis using apriori association rule. *International Journal of Pure and Applied Mathematics, 119*(7).

Surjandari, I., & Seruni, A. C. (2005). Design of product layout in retail shop using market basket analysis. *MakaraTeknolgi, 9*(2), 43–47.

Szymkowiak, M., Klimanek, T., & Jozefowski, T. (2018). Applying market basket analysis to official statistical data. *Econometrics Ekonometria Advances in Applied Data Science, 22*(1), 39–57.

Tan, P.-N., Steinbach, M., & Kumar, V. (2016). *Introduction to data mining*, pp. 2–3. ISBN 978-93-3257-140-2.

Tatiana, K., & Mikhail, M. (2018). Market basket analysis of heterogeneous data sources for recommendation system improvement. *Procedia Computer Science*, 246–254. ISSN 1877-0509.

Improving Blast Furnace Operations Through Advanced Analytics

Rishabh Agrawal and R. P. Suresh

1 Introduction

The blast furnace is a huge stack where the iron oxides are chemically reduced and physically converted to liquid iron called hot metal. The furnace is lined with refractory brick where iron ore, limestone and coke are fed from the top and hot blast (preheated air) is blown up from the bottom. The high-temperature air blowing up reacts with coke to constitute a combustion and releases heat which releases the oxides from iron ore and reduces the iron oxide to molten iron. The molten iron is drained from the furnace bottoms into torpedo which carries the hot metal-to-basic oxygen converter to convert the hot metal to steel (see Ricketts).

The full blast furnace assembly constitutes not only the blast furnace but also a heat recovery section which is commonly referred as hot blast stoves. The hot blast stove supplies hot air to the blast furnace. The air from atmosphere is heated inside the stoves using blast furnace top gas in cycles. The preheated air coming out of stoves is called hot blast which is delivered to the blast furnace through tuyeres. In this way, the energy coming out from the top of blast furnace as top gas is recovered back to the blast furnace in terms of hot blast. The higher the temperature of the hot blast, lower will be the fuel–coke/coal requirement in the blast furnace. Thus, the good recovery in this heat recovery section directly reduces the cost of hot metal and thus steel (see Satyendra 2015).

R. Agrawal (✉) · R. P. Suresh
Supply Chain and Operations Analytics, Applied Intelligence, Accenture Digital, Gurugram, India
e-mail: rishabh.c.agrawal@accenture.com

A. K. Laha (ed.), *Applied Advanced Analytics*, Springer Proceedings in Business and Economics, https://doi.org/10.1007/978-981-33-6656-5_10

1.1 Literature Review

Wang (2016) presented some work on the oxygen enrichment to hot stoves. The stoves were heated by the blast furnace top gas coming out of blast furnace top. The paper presented that oxygen enrichment in the stoves can improve the performance of the stove. This can be done by increasing the dome temperature, etc. Wang says "CFD (Computational flow dynamics) modeling work indicated that the oxygen lance position is one important factor to achieve a uniform mixture of oxygen and combustion air for the combustion process."

Zetterholm (2015) developed a dynamic model where thermophysical properties of both gas and solid were calculated with respect to time and position in the stove. He says that the model could be used to represent the stove. Oxygen enrichment was studied in the paper as a major factor to improve the stove operation and in turn produce hot blast with higher temperature to improve the efficiency of the blast furnace.

Simkin (2015) developed Mamdani fuzzy control model to reduce the consumption of the natural gas in heating the stoves. The paper says that "One of the main advantages of fuzzy knowledge base is the ability to use minimum information about the modeled object." It is formed on the information about the input and the output parameters of the stove. It takes into account the heat of combustion going into the stoves or the total calories of the fuel used to heat the stoves. This takes care of the problems that arise due to control systems or nonlinear functions related to the blast air.

Butkarev (2015) presented that the hot blast temperature can be increased by 30–40 °C by improving the automatic control system designed by OAO VNIIMT.

Lin (2007) discussed that on studying the effect of fuel gas preheating on the thermal efficiency of the hot blast generating system, experimental results revealed that the efficiency of the hot blast generating process was increased from 75.6% to 78.7%. According to the test results, "the advantage of operating the heat recovery system was evident in the reduction of fuel gas depletion rate by 822 L/h on an oil equivalent basis. The annual energy saving over 1.9 million US dollars and the CO_2 reduction 15,000 ton/year can be achieved."

1.2 Purpose of Research

The current state of stove operations is not standardized. Many times, operators based on their experience take critical decisions. To standardize the decision-making process in the most optimum way, an analytics research project with one of the Asia's largest iron and steel manufacturing plants was kick-started. After examining the available data, it was observed that the hot blast temperature varied between 1175 °C (10th percentile) and 1204 °C (90th percentile) between January 1, 2018, and September 30, 2018. The median hot blast temperature was 1192 °C.

The purpose of analytics research is to enable the plant operations to operate the blast furnace stoves in a way that it maximizes the energy recovery in the stoves, thus increases the hot blast temperature entering the blast furnace and hence reduces the fuel requirement inside blast furnace. The end result of the research would be to reduce the variability and increase the median temperature, eventually reducing the cost of hot metal.

2 Data Availability

We used the data collected from a steel manufacturing process. The data is captured through the SCADA system (Supervisory Control and Data Acquisition) which is often called the Level-1 data with almost zero process lag. The Level-2 data with <1 min lag is present through the process historians. The dataset identified to use for the project is present at 1 min. level granularity, and 8.5 months of plant operation data was available which included controllable and non-controllable variables (see Fig. 1). Variables are as follows: hydrogen%, CO%, CO_2% in blast furnace top gas, blast furnace top gas flow, blast furnace top gas temperature, combustion air flow to stoves, combustion air temperature to stoves, wind volume, cold blast temperature, hot blast temperature, stove dome temperature, stove refractory temperature, stove exhaust gas temperature and cold blast flow. Please refer below figure to understand the variables (see Fig. 2).

2.1 Understanding the Present Control System

There are 4 stoves in parallel operation to heat up the blast entering the blast furnace. Stoves are operated in cycle. First, a stove is heated up to certain temperature and

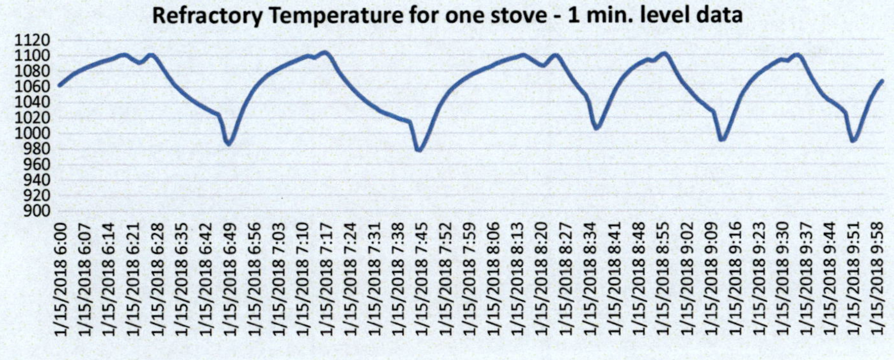

Fig. 1 4 h—minute-level data for refractory temperature

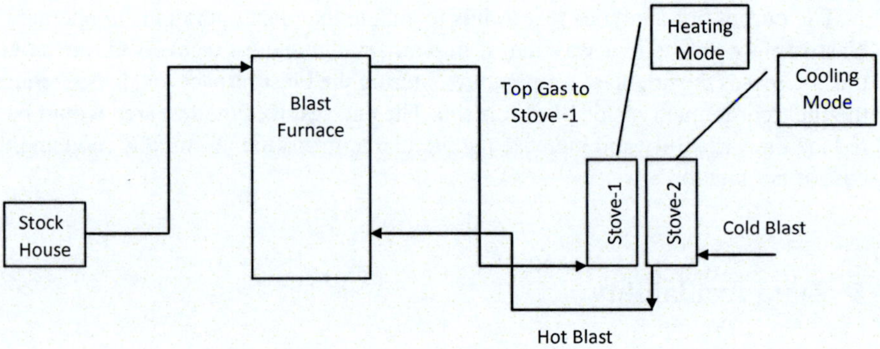

Fig. 2 Blast furnace process diagram—overview

then cold blast at ~200 °C is heated by passing through the heated stove to ~1200 °C. In the process of heating the cold blast, the stove's temperature comes down, it again is heated and hence the cycle continues. Hence, the cycle is divided into two parts: heating period and cooling period (Fig. 3).

The control system during the heating and soaking period includes dome temperature, exhaust temperature and air-to-gas ratio. While the refractory temperature steadily rises as soon as the heating starts, dome temperature on the other hand rises very steeply and reaches 1300 °C in between 5 and 10 min. The upper limit of dome temperature is set at 1350 °C after which the cycle stops. The exhaust temperature measured near the exit of the hot flue gases from stove bottom also steadily rises and is set at a high limit of 350 °C after which the cycle stops.

Once the dome temperature hits 1300 °C, the heating cycle is over, and soaking cycle starts. In the heating cycle, only one air-to-gas ratio ($A1$) is operational which is controlling the flow to combustion air to the stove. Once, the soaking cycle starts, the 2nd air-to-gas ratio ($A2$) also becomes active and the control system varied the air flow to stoves between the $A1$ and $A2$ ratios to keep the dome temperature below

Fig. 3 Heating, soaking and cooling cycle

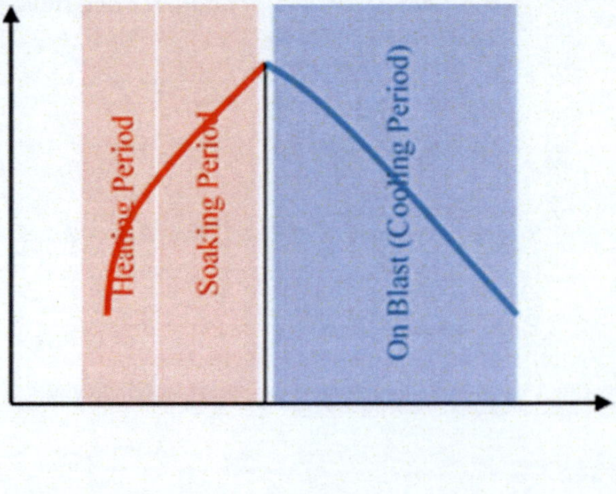

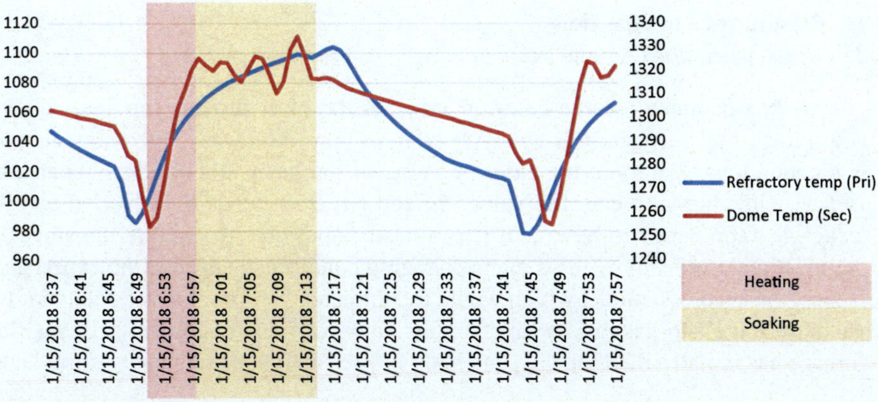

Fig. 4 Dome temperature and refractory temperature curves

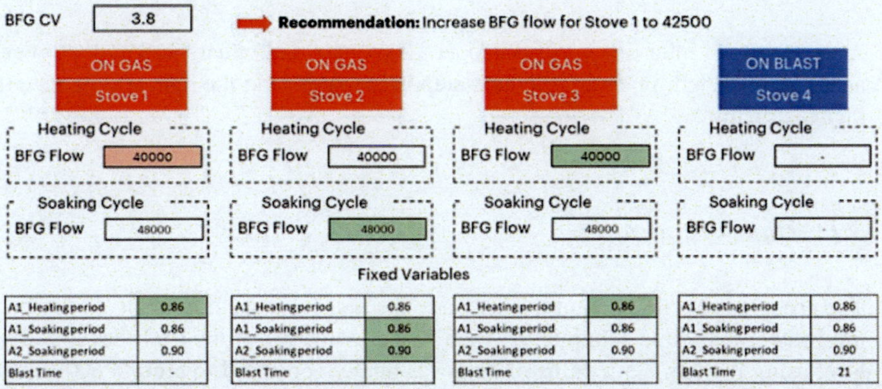

Fig. 5 Dashboard

1350 °C so that the cycle does not stop prematurely and maximum possible refractory temperature is reached. As the air-to-gas ratio increases, the exhaust temperature increases, and as the air-to-gas ratio is reduced, the dome temperature increases. The $A1$ and $A2$ ratios are set by the operators at the start of the heating cycle, and thus determining optimum $A1$ and $A2$ for different stoves is important to maximize the hot blast temperature.

After studying the plant control system, it was understood that there is no control during the cooling period. It is the correct handling of control system during the heating cycle which would yield a longer cooling cycle (longer implies more heat transfer to cold blast). Each individual heating cycle and cooling cycle are typically 30 min long.

During the heating period of the stoves, there are two controlling parameters which the operators could handle and manipulate for effective heating of the stoves.

1. Blast furnace top gas flow
2. Air-to-gas ratios (A1 and A2).

One key parameter is the calorific value of the blast furnace top gas which is determined by hydrogen, CO and CO_2 composition in the gas. Since blast furnace is a continuous operation, the calorific value of top gas is continuously changing. Hence, with changing calorific value, the top gas flow needs to be varied and the air-to-gas ratio needs to be tuned. It is almost impossible to operate the stoves in the most optimum way with so many changing conditions with just the experience. Hence, the need of a model which would recommend the optimum set points for top gas flow and air-to-gas ratio with changing calorific value of top gas arises. (Blast furnace gas is also called top gas, and both will be interchangeably used from here.)

3 Data Analysis and Modeling

After rounds of interviews with the operations team and understanding the process and control system in detail, the next step is to understand the data and finalize the model structure.

3.1 Data Preparation

The process is cyclic in nature. Therefore, the analysis would be done on the cycle level rather than time stamp level. Data was transformed from 1 min level to cycle level using Python. 8.5 months of data was finally converted to around 6000 cycles for one stove. The periods in which the plant was under transition state/unsteady were removed (see Fig. 4).

The red zone above indicates the heating cycle. The heating cycle is demarcated by the continuous increase in the dome temperature—when the temperature stops rising, the end of the heating cycle is indicated. During the soaking cycle, the dome temperature is seen to be sinusoidal in nature. The end of the soaking cycle is indicated when the refractory temperature starts decreasing. In this way, the minute-level data was converted to cycle-level data.

3.2 Modeling

To standardize the process and maximize heat recovery, it is imperative to understand how the cycles which yielded high hot blast temperature were operated. Unfortunately, hot blast temperature is a function of all the 4 stoves' operation; hence, best cycle identification based on hot blast temperature would yield faulty results. Some descriptive analytics to understand the relation of refractory temperature of each stove with the hot blast temperature was done. Refractory temperature was divided into deciles, and average hot blast temperature in each decile was plotted. It was evident that hot blast temperature is a function of each stove refractory temperature. Therefore, identifying best cycles based on refractory temperature would be correct as refractory temperature of one stove would have no correlation with any other stove operation. Hence, maximizing refractory temperature during the heating cycle becomes key to achieving the goal of this study.

Taking all these exploratory analyses into consideration, unsupervised machine learning approach was used to understand the best plant operations. K-means clustering algorithm was used to model and divide the clusters based on dome temperature, exhaust temperature, combustion air flow during heating period, combustion air flow during soaking period, blast furnace gas flow during heating period, blast furnace gas flow during soaking period, calorific value of blast furnace gas during heating period, calorific value of blast furnace gas during soaking period, heating time, soaking time, blast furnace gas temperature and combustion air temperature variables for one stove at a time.

Scree plot with number principal components on x-axis and eigenvalue of principal components on y-axis were plotted. The elbow formation was observed to be at three principal components for Stove 1, three for Stove 2, five for Stove 3 and three for Stove 4. Taking reference from these plots, the number of clusters was decided.

3.3 Model Validation

The model was developed in Python using scikit-learn library. The cluster validation was performed based on multiple variables. The first validation was to see if meaningful clustering has taken place based on refractory temperature. It was observed that refractory temperature in different clusters followed distribution tabulated below (see Table 1):

As the stoves are heated from blast furnace/top gas, the composition of which is not controllable, it is imperative to check if the clusters formed in the above table have similar distribution of the calorific value of the top gas. This is to validate that the clusters are not modeled on the basis of calorific values as that cannot be the basis for recommendation.

The best cluster must give best refractory temperature for all the calorific value ranges. The calorific value was divided into deciles, and average refractory temperature for best cluster was compared with average refractory temperature of the other clusters. It was found that for all calorific ranges the best cluster had maximum refractory temperature.

The best cluster's air-to-gas ratio and blast furnace gas flow for different calorific values were shared with the client to make the operators follow the model recommendations.

Table 1 Cluster results for all stoves

Stove#	Cluster#	25th percentile refractory temperature	50th percentile refractory temperature	75th percentile refractory temperature
Stove 1	Cluster1	1075.5	1081.7	1090.5
	Cluster2	1036.4	1071.9	1086.7
	Cluster3	1032.0	1071.6	1083.2
Stove 2	Cluster1	1070.2	1080.0	1083.4
	Cluster2	1067.8	1072.9	1078.6
	Cluster3	991.9	1058.1	1071.4
	Cluster4	1045.5	1060.2	1069.7
	Cluster5	1047.6	1060.8	1070.0
Stove 3	Cluster1	1094.6	1104.7	1113.6
	Cluster2	1100.8	1111.5	1119.9
	Cluster3	1114.6	1120.9	1126.2
Stove 4	Cluster1	1013.4	1044.4	1049.3
	Cluster2	1021.0	1034.6	1044.1
	Cluster3	1025.4	1040.5	1047.6

Web-based live dashboard was designed on .NET framework which takes live data from data server and updates the recommendations on real-time basis. Operator control room training was organized to train the operators to use the dashboard. Please refer to sample dashboard below (see Fig. 5).

To improve the usage of the model and to evaluate the model performance, it is very important to calculate the extent of compliance of the operators to the model recommendations. The compliance was calculated using Python script where the recommended BFG flow and $A1$ and $A2$ ratios are compared with the actual BFG flows and $A1$ and $A2$ ratios. The margin of error has been kept at 2.5%. If the actual numbers fall in that window of recommended numbers, the operators are called to be compliant.

The model is set to be retuned in every 1 month as the efficiency of the process equipment related to the process keeps changing with any change in the process. The model tuning is done by replacing the oldest data of the 8.5-month window with the latest one-month data to include any recent change in the process. However, if any major turnaround/shutdown maintenance activity takes place, which includes change in any equipment, then complete model data overhaul needs to be done.

4 Results and Conclusion

Hot trials were carried out for a week. The median hot blast temperature increased to 1201 °C, 10th percentile temperature increased to 1188 °C and 90th percentile

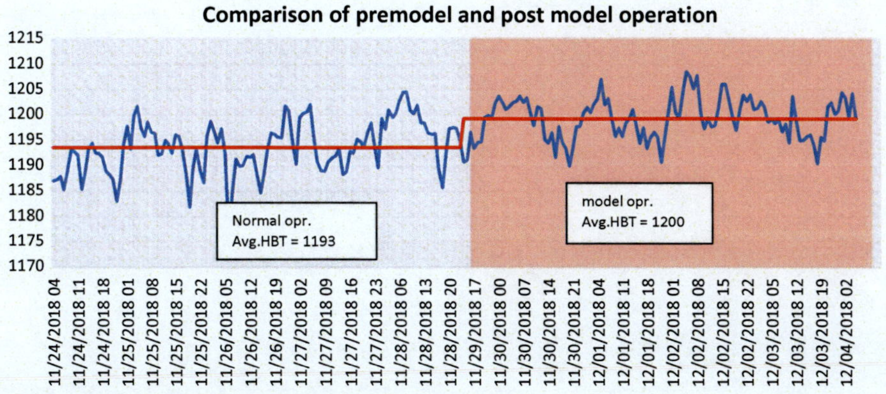

Fig. 6 Results

temperature increased to 1208 °C. Hence, the results indicate improvement in variability as well as the median temperature (see Fig. 6).

The blast furnace operations run according to the model generated recommendations improved the hot blast temperature, hence reducing the cost of the hot metal.

References

Butkarev, A.A. (2015). Boosting the hot-blast temperature in blast furnaces by means of an optimal control system. Stal No. 3.

Lin, P.-H. (2007). *Efficiency improvement of the hot blast generating system by waste heat recovery.* Kaohsiung, Taiwan: Energy and Air Pollution Control Section, New Materials Research and Development Department, China Steel Corporation.

Ricketts, J. A. How it works: The blast furnace. https://www.thepotteries.org/shelton/blast_furnace.htm.

Satyendra. (2015). Generation of hot air blast and hot blast stoves. https://ispatguru.com/generation-of-hot-air-blast-and-hot-blast-stoves/.

Simkin, A. (2015). Control model of the heating hot blast stove regenerative chamber based on fuzzy knowledge with training set. Metallurgical and Mining Industry.

Wang, C. (2016). Modelling and analysis of oxygen enrichment to hot stoves. In *The 8th international conference on applied energy—ICAE2016.*

Zetterholm, J. (2015). Model development of a blast furnace stove. In *The 7th international conference on applied energy—ICAE2015.*

Food Index Forecasting

Kalyani Dacha, Ramya Cherukupalli, and Abir Sinha

1 Business Problem

Time series forecasting techniques are important to improve business process, increase efficiency, profits and reduce costs. Forecasting in any business is important because it provides an insight about the direction and the volume business is heading towards.

In this paper, we are discussing business problems of one of the leading food service providers in North America which operates in industries like education, health care, sport and entertainment, and business and government. They do not buy live animals (e.g., hogs, cattle, poultries) to make the finished product. Rather, they buy raw products and make intermediate products out of them (e.g., ground beef, steak, bacon, liquid eggs) from several food vendors. The company signs long term contracts with the food vendors either for 12 months or for 18 months in advance. Hence, it becomes critical to understand where the inflation numbers for each individual food category (U.S Bureau of Labor Statistics, BLS published non-adjusted numbers) are heading toward. Currently, they guess the average inflation number for the upcoming 12 or 18 months for each individual category, adjust their last year inflation numbers in the contract and renew it. So, to make the contract efficient and profitable for them, it is necessary to estimate the upcoming average inflation numbers effectively. We were provided with a list of 18 major food categories that contribute to 90% of their food supply business.

K. Dacha (✉) · R. Cherukupalli · A. Sinha
Deloitte Consulting, Hyderabad, India
e-mail: kalyani.dacha@gmail.com

R. Cherukupalli
e-mail: ramyac1@gmail.com

A. Sinha
e-mail: abirsinha1@gmail.com

A. K. Laha (ed.), *Applied Advanced Analytics*, Springer Proceedings in Business and Economics, https://doi.org/10.1007/978-981-33-6656-5_11

Table 1 List of product categories

S. No.	Category	Sub-category
1	Beef	Ground beef
2		Beef roast
3		Beef steak
4	Pork	Bacon
5		Ham
6		Pork chops
7	Poultry	Chicken
8		Fresh and frozen chicken
9		Turkey
10	Eggs	Eggs
11	Potatoes	Potato
12		Frozen and freeze-dried prepared foods
13		Potato fries (PPI)
14	Fish and seafood	Fish and sea food
15	Dairy	Milk
16		Cheese
17		Coffee
18	Fats and oils	Fats and oils

Price index forecasting was used to forecast monthly inflation numbers for 18 product categories namely (Table 1):

2 Data Gathering

All the information used for this analysis has been downloaded using publicly available, free, external websites.

Data gathering was primarily carried out for the dependent variable (Consumer Price index). BLS website stores monthly price index/inflation numbers for over 100 food items, and updates are carried out each month. Extensive research was carried out to find out what can drive price for each product category. Instead of looking into the individual sub-categories (like ground beef, beef roast and beef steak)—research was focused on the overall categories (like what affects beef prices in this example).

We can broadly categorize key drivers across major product categories into five different levels (Fig. 1):

Few insights from the research on affecting price trends:

1. It was observed that feed expenses were high in 2012, and huge number of cows were slaughtered (Frohlich 2015) as an immediate impact. In the next

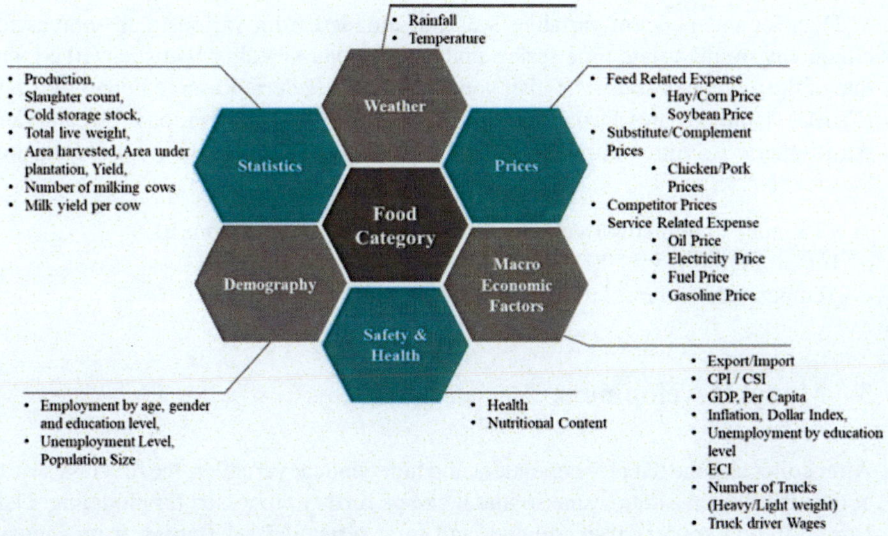

Fig. 1 Key drivers across food categories

two years, number of meat producing cows were quite less in number, and it significantly affected beef price in 2014. Hence, feed prices (hay, corn, soybean) were identified as one of the key drivers

2. Butterfat percentage: Milk (Bailey 2017; Cushnahan 2003) butterfat content lowest at peak production and highest toward the end of lactation
3. Biofuels (Rosillo-Calle et al. 2009; Parcell et al. 2018),[1] soaps, washing powders, personal care products have close interdependence to oilseed and biodiesel markets. Any changes in either US biodiesel policy or global biodiesel policy could shock oilseed prices
4. El Niño[2] is an abnormal weather pattern caused by the warming of the Pacific Ocean near the equator, off the coast of South America. In South America, there is a drastic increase in the risk of flooding on the western coast, while there is an increase in the risk of droughts on parts of the eastern coast. In eastern countries, like India and Indonesia, there is an increase in droughts. These affect the fish and seafood (OECD/FAO 2016) price index a lot.
5. The main drivers for decline of price of the commodities will be the competitive prices of substitutes (like eggs, chicken, etc.) the slowdown in demand from key markets due to sluggish economic growth and reduced production and marketing costs of aquaculture products due to lower transport and feed costs
6. Political situations in Brazil, Indonesia affects the coffee (van den Brom 2020)[3] price most as they are the leading coffee producers.

[1] An overview of the Edible Oil Markets: Crude Palm Oil vs Soybean Oil" (July 2010).

[2] Rinkesh, "What is El Niño?".

[3] Jack Purr, "What affects the price of coffee.

Data for independent variables was downloaded from various data sources for conducting multivariate time series analysis. Bureau of Labor Statistics (BLS) was one of the main data sources. Other sources include Federal Reserve Economic Data (FRED), United States Department of Agriculture (USDA), National Oceanic and Atmospheric Administration (NOAA), Data World, National Agricultural Statistics Service (NASS).

- o CPI index and unemployment rate were mainly sourced from BLS,
- o Import/export from USDA,
- o Temperature and rainfall data from NOAA.

3 Model Development

After collecting the data for dependent and independent variables, the first task was to collate the data in a data frame so that it can be further processed for modeling. Data from different sources was collated, and time series dataset starting from January 2009 was created. The area of interest in this study was to model the Year-over-Year (Y–o-Y) inflation numbers for all the dependent variable categories. Due to the volatile nature of Y–o-Y variable, modeling was done on the actual index data for all the food categories, and then finally, the forecasted numbers were converted into Y–o-Y for further analysis.

3.1 Pre-modeling

Given this is a time series dataset, it is highly likely that lagged version of independent variables might influence the dependent variables the most. Given we are trying to forecast for future months, lagged independent variables can be directly used (given data is available), else we can also forecast directly. After the creation of lags for each of the independent variable, correlation with the dependent variables was calculated for all the 13 versions of each independent variable, i.e., original variable (Lag 0) and its 12 lagged versions. For example,

- For Ground Beef Price Index, petrol price (six months' lag) correlated the most.
- Beef slaughter count lag two variable was the highest correlated variable with Ground Beef Index with correlation −0.78.
- Lag 1 of shell egg import 1000 dozen is the highest correlated variable with Egg Index with correlation 0.34.

Data considered for further analysis was in the period Jan 2010–May 2018, and there were no missing values within this time frame. For providing price index forecasts, five modeling techniques (univariate and multivariate) were applied. Best model was chosen from the five techniques based on the applicability, accuracy and

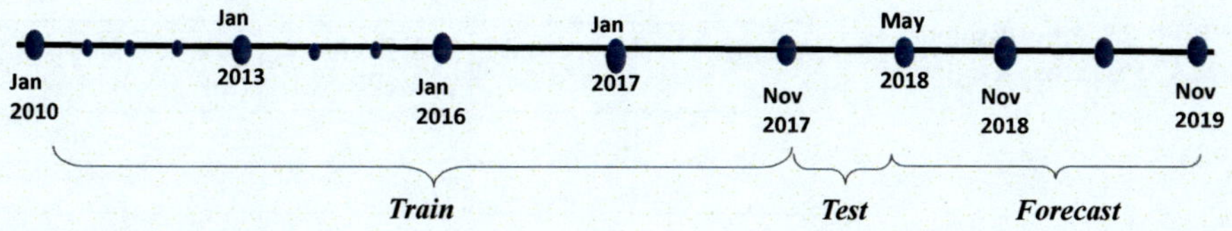

Fig. 2 Modeling timeframe

blind validation. Forward forecast inflation (average and interval range) for 12 and 18 months is provided as results.

3.2 Modeling Process

After identifying all the highly correlated variables for each of the 18 dependent variables separately, next task was to run models to identify the significant variables for each of the dependent variable category.

Train, test and forecast periods were created as given in Fig. 2 throughout the modeling process:

To define train and test period of the study, several test periods were taken into consideration like: 12 months, 9 months and 6 months, respectively. While observing the pattern of the test period (12, 9 and 6 months), it was seen that for most of the categories, distribution of the test set was completely different from that of train period, and as a result, the forecasts were going in a completely different direction. To avoid this situation, test periods were reduced to six months to better train the time series models. To compare the performance of various models, **MAAPE** was used. Below is the definition and reasoning behind using this metric.

3.3 Evaluation Metrics

In time series analysis, mean absolute percentage error (MAPE) is widely used and is calculated as below.

$$\text{MAPE} = \frac{100\%}{n} \sum \frac{|\text{Actual} - \text{Forecast}|}{|\text{Actual}|}$$

In this case study, Y–o-Y is the dependent variable, and this can take both positive and negative numbers, and using MAPE as a model evaluation criterion was found to be not applicable as errors were huge. Example is shown below:

Table 2 Comparison of MAAPE across test periods

Model #	Test period	YoY Train MAAPE (%)	YoY Test MAAPE (%)
1	12	18	83
2	9	21	40
3	6	20	**22**

- Assume for the forecast month June 2018, for the best model, the estimate is 2.50, whereas the actual number is 0.50. In this case, the MAPE would be |0.50 – 2.50|*100/|0.50| = 400% which is not reflecting the actual scenario.
- Hence, alternative model evaluation criteria MAAPE (Kim and Kim 2016) was used as it is applicable to deal with positive or negative numbers where as MAPE was not. MAAPE is calculated using the below formula:

$$\text{MAAPE} = \frac{100\%}{n} \sum \arctan\left(\frac{|\text{Actual} - \text{Forecast}|}{|\text{Actual}|}\right)$$

- For the above example, MAAPE is 132%. Another advantage is MAAPE is well defined even if Y–o-Y is zero though MAPE is not.
- Table 2 describes different test periods and the corresponding MAAPE for one of the food categories index.

The modeling techniques used in this paper is given in Table 3.

- Univariate models were run on the training set, i.e., directly considering the dependent variables as time series. Then, the fitted models were used to forecast the test period to check for the accuracy of the model. Accuracy was calculated using the six original test data points versus six forecasted data points. Once the model was finalized from the accuracy metric, model parameters were retrieved from the best model, retrain the model using the same set of parameters, but with the data of both train and test and finally forecast for the final 18 months. For example, for chicken fresh and frozen category—say ARIMA parameters $p = 2$, $d = 1$, $q = 0$ was finalized. Using the parameters, forecast for the upcoming 18 months will be developed using ARIMA parameters $p = 2$, $d = 1$, $q = 0$ on the entire train + test (95 + 6 = 101 months) data.

Table 3 Modeling techniques used

Method	Technique
Univariate	ARIMA (Box and Jenkins 1970)
Univariate	Holt-Winters (Chatfield and Yar 1988)
Univariate	Exponential smoothing (Broze and Mélard 1990)
Multivariate	Regression (Ramcharan 2006)
Multivariate	ARIMAX (YuanZheng and Yajing 2007)

- For multivariate time series models, independent variables were forecasted first to get the final forecast numbers for the dependent variables. For each of the dependent categories, regression models were run using independent variables in the training period. Once the model variables were finalized, next step was to forecast for the test period. This was done in a two step processes. First was to forecast for the independent variable, and once this was complete, next step was to get the forecast numbers for the dependent variable from the regression equation. Forecast of the independent variables was done using the three univariate time series methods (listed in Table 3), and selected method was the one for which the test MAAPE was the least. Once all the forecast numbers were handy for all the independent variables, regression equation was used to get the final forecast numbers for the dependent variables.
- For ARIMAX model, same set of independent variables and their corresponding forecasted numbers were used. To select the optimal parameters of the ARIMAX model, grid search was carried out.

4 Results

Out of the 18 models built for the study, we have achieved greater accuracy (<35% MAAPE) for 50% of the models, and error range for the rest of the models was between 40 and 60%.

Among the 18 food categories, majority were stable (Fig. 3) with the overall Y–o-Y range between ±5%, few categories like eggs (Fig. 4) are a volatile category with the Y–o-Y range between ±40%, however, our models were robust, and we achieved great accuracy in such scenarios too.

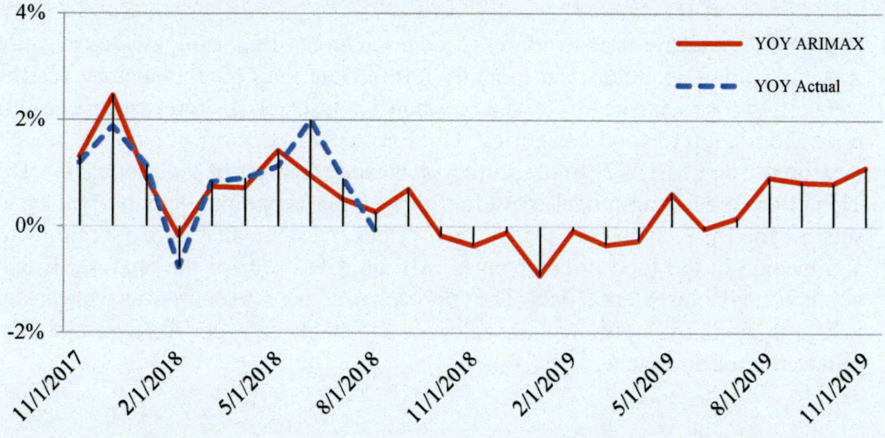

Fig. 3 Figure showing Y–o-Y plots comparison between Actual (Historical Cheese values shown as blue dotted line) versus ARIMAX (forecasted values based on the model selected based on performance as red line)

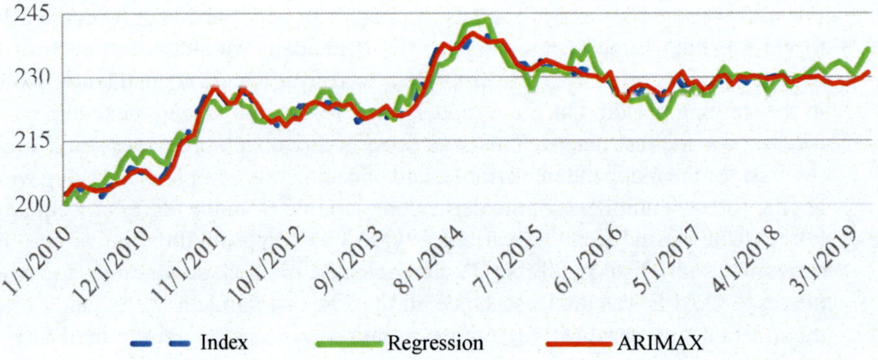

Fig. 4 Price index comparison between Actual (Historical Cheese values shown as blue dotted line) versus ARIMAX (forecast values based on ARIMAX model as red line) versus Regression (forecast values based on regression model as green

Below is the summary report of 18 models built with details on final model selected (based on performance); MAAPE metrics for train and test periods are provided in Table 4. ARIMAX had been predicting better than rest of the models in most categories. Beef category predictions were 70–80% accurate, and these were the main categories. Of all categories, dairy and poultry are having greater accuracy. Except for pork and potatoes category, rest all category forecasts were in the range of 60–80%. Pork and potatoes categories were highly volatile, and the greatest accuracy achieved in these cases is about 50%.

5 Conclusion

In this study, we have tried to address food price forecasting using various univariate and multivariate techniques at monthly level. More than 60 explanatory variables were tested for each category based on extensive research for forecasting consumer price indexes of 18 food categories. The forecasting performance of the model is measured using MAAPE, and accuracy achieved for most of the models is <15%. This price-forecasting model is useful in capturing economic demand-pull factors such as food use, substitute prices, feed prices, weather, macro-economic factors and income in the food price changes. All the data used for the analysis is using publicly available, external data. The approach used here is robust—as we were able to capture trend for highly volatile category and stable category likewise and obtain satisfactory performance.

Table 4 Detailed results for all the categories

Model #	Category	Sub-category	Final model	Train MAAPE (%)	Test MAAPE (%)
1	Beef	Ground beef	ARIMAX	20.32	22.31
2		Beef roast (round)	ARIMAX	25.19	34.29
3		Beef steak	ARIMA	27.66	46.24
4	Pork	Pork bacon	ARIMAX	34.22	58.42
5		Pork ham	ARIMAX	29.45	53.04
6		Pork chops	ARIMAX	33.16	46.78
7	Poultry	Chicken	ARIMAX	28.02	28.32
8		Fresh and frozen chicken parts	ARIMA	36.66	25.98
9		Turkey	ARIMAX	40.42	36.34
10	Eggs	Eggs	ARIMAX	36.72	35.37
11	Potatoes	Potatoes	ARIMAX	37.55	111.84
12		Frozen and freeze-dried prepared foods	ARIMAX	23.28	59.62
13		Potato fries [PPI]*	ARIMAX	46.49	16.23
14	Fish and seafood	Fish and sea food	ARIMAX	28.02	34.28
15	Dairy	Milk	ARIMAX	27.09	25.20
16	Coffee	Coffee	ETS	27.68	31.74
17	Dairy	Cheese	ARIMAX	26.77	28.65
18	Fats and oils	Fats and oils	ARIMA	33.19	43.89

References

Bailey, H. (2017, September). Dairy risk-management education: Factors that affect U.S farm-gate milk prices.

Box, G. E. P., & Jenkins, G. M. (1970). *Time series analysis: Forecasting and control*. San Francisco: Holden-Day Inc.

Broze, L., & Mélard, G. (1990). Exponential smoothing: Estimation by maximum likelihood. *Journal of Forecasting, 9,* 445–455.

Chatfield, C., & Yar, M. (1988). Holt-winters forecasting: Some practical issues. *Journal of the Royal Statistical Society. Series D (The Statistician), 37*, 129–140.

Cushnahan, A. (2003, April). Factors influencing milk butterfat concentration.

Frohlich, T. C. (2015, April). States killing the most animals for food. https://www.usatoday.com/story/money/business/2015/04/15/247-wall-st-states-killing-animals/25807125/.

Kim, S., & Kim, H. (2016, September). A new metric of absolute percentage error for intermittent demand forecasts.

OECD/FAO. (2016). Fish and seafood. In *OECD-FAO Agricultural Outlook 2016–2025*.

Parcell, J., Kojima, Y., Roach, A., & Cain W. (2018, January). Global edible vegetable oil market trends.

Ramcharan, R. (2006). Regressions: Why are economists obessessed with them? Accessed 2011–12–03.

Rosillo-Calle, F., Pelkmans, l., & Walter, A. (2009, June). A global overview of vegetable oils, with reference to biodiesel.

van den Brom, J. (2020). Coffee.

YuanZheng, W., & Yajing, X. (2007). Application of multi-variate stable time series model ARIMAX. *Statistics and Operation, 9B*, 132–134

Implementing Learning Analytic Tools in Predicting Students' Performance in a Business School

R. Sujatha and B. Uma Maheswari

1 Introduction

Developments in the field of information technology with respect to big data have resulted in disruptive implications across all sectors (Baradwaj and Pal 2011). Data is now available in abundance, and therefore, there is a need to employ tools and techniques to mine such data. Data mining tools and techniques have found applications across various disciplines including customer profiling, fraud detection, DNA sequencing, etc. (Lauria and Baron 2011). Educational data mining (EDM) and learning analytics (LA) are two communities evincing a keen interest in how big data could be exploited for the larger benefit of the education sector (Baker and Inventado 2014). EDM deals with "developing methods that discover knowledge from the data originating from educational environments" (Han and Kamber 2006). Such mining techniques result in pattern recognition which forms the basis for decision making and support interventions (Siemens et al. 2011) ultimately leading to optimizing the learning process (Baker and Siemens 2014).

LA on the other hand emphasizes more on the data visualization and human intervention and has evolved into a critical domain in the education space (Gasevic et al. 2016). LA is defined as "the measurement, collection, analysis and reporting of data about learners and their context, for purposes of understanding and optimizing learning and the environment which it occurs" (Ferguson 2012; Elbadrawy et al. 2016). Research in the field of EDM and LA has clearly demonstrated the need for understanding the teaching–learning process and using this information for improving the same (Baker and Inventado 2014; Gasevic et al. 2016).

R. Sujatha (✉) · B. Uma Maheswari
PSG Institute of Management, Coimbatore, India
e-mail: sujatha@psgim.ac.in

B. Uma Maheswari
e-mail: uma@psgim.ac.in

A. K. Laha (ed.), *Applied Advanced Analytics*, Springer Proceedings in Business and Economics, https://doi.org/10.1007/978-981-33-6656-5_12

2 Purpose of Research

A solid theoretical framework in this domain has not evolved yet. The vast differences in the kind of learning tools and educational systems in different institutions across developed versus developing nations clearly delineate the impracticality of a model which could be generalized across geographies as well as across disciplines (Agudo-Peregrina et al. 2014). Studies clearly proved that self-regulation of learning (Black and Deci 2000), self-efficacy (Chung et al. 2002) and information-seeking behavior (Whitmire 2002) vary with course and discipline. Early identification of students who are likely to fail in a particular subject gives scope for early interventions and therefore helps faculty to provide more attention to a specific set of students.

The review of literature showed that existing studies in this domain has been based on data extracted from learning management system (LMS) (Romero et al. 2013). One such LA model was developed at Purdue University called Course Signals (Arnold and Pistilli 2012) which successfully transformed a research agenda into a practical application. LMS data was used in pattern recognition of user behavior (Talavera and Gaudioso 2004). Studies have also been undertaken with data obtained from Moodle (an open-source course management system) (Romero et al. 2013). Demographic characteristics and course management system usage data was used to develop predictive machine learning models (Campbell 2007). Unstructured data obtained from online discussion forums was used to perform sentiment analysis in a study conducted by Laurie and Timothy (2005). Lykourentzou et al. (2009) used the scores of a quiz activity to cluster the students using neural networks.

Applying analytics in education is the need of the hour, especially in the context of a developing economy like India. The inferences drawn from prior studies have been eagerly accepted by the academic community (Gasevic et al. 2016). Hence, it is time for educational institutions to use machine learning tools to enhance teaching–learning experience. This study deploys learning analytics technique using the data of students undergoing a post-graduate management program and attempts to create a system of preventive feedback mechanism for faculty and students.

3 Research Objectives

The objective of the study was to create an early intervention mechanism to enhance students' performance. This study aims to develop a predictive model using machine learning algorithms to predict the academic risk of a student passing or failing a course (binary response) and the marks of the student (continuous data) in a course based on past data. The research objectives of the study are

o RO1: To develop a model to predict the academic status of a student in a course?
o RO2: To develop a model to predict the grade of a student in the course?

4 Methodology

The learning analytics model adopted in this study is based on supervised learning algorithm. The training data set had both predictive features including demographic characteristics of the students, as well as the response feature which is the students' academic status. The methodological framework developed by Lauria and Baron (2011) has been adopted in this study. The framework includes five steps such as data collection, data preparation, data partition, building the models and evaluating the models.

4.1 Data Collection

The study was conducted among post-graduate management students undergoing the Master of Business Administration (MBA) program. The data was collected during the admission and during the progression of the course. Demographic data of the students was collected from the admissions portal and the academic performance of the students from the examination portal. The academic performance of the students relating to six foundation courses undertaken by the student in the first semester and one capstone course strategic management was considered for the purpose of this study. Past three years data was considered ($n = 522$).

4.2 Data Preparation

The data was pre-processed for missing values, outliers and incomplete data. Few students' data who left the college was identified and removed from the database. Next step, the identity of the students represented by names and roll numbers was removed to ensure anonymity. A few features had to be derived for the purpose of this study. The data relating to the 'date of birth' of the students was extracted from the portal, and the age of the student at the time of joining the course was derived. The other derived feature was 'break in study.' The data relating to the undergraduate degree completion year was extracted. This data showed that most of the students did not have a break in study; therefore, this feature was dichotomized into 'break in study,' Yes/No. Data transformation was done in the case of four features (community, tenth board, higher secondary board, undergraduate degree). The feature 'community' had seven categories such as OC, BC, MBC, DNC, SC, ST and Others. A summary of the feature showed a small percentage of the students belonged to MBC, DNC, SC and ST. Therefore, these categories were combined with 'Others' category resulting in only three categories of 'community,' namely OC, BC and Others. The tenth and the higher secondary board also went through a similar process to result in four categories, namely Tamil Nadu Board, Kerala Board,

CBSE and Others. As the MBA program did not have any restrictions in terms of undergraduate degree, this program attracts students from versatile disciplines. This feature also had to be transformed into three major groups. namely Arts, Engineering and Science. The features used for model building is shown in Table 1.

Table 1 Features in input dataset

S. No.	Features	Continuous/categorical	Type of features
1	Age	Continuous	Numeric
2	Gender	Categorical	Male/Female
3	Community	Categorical	OC/BC/Others
4	Tenth Percentage	Continuous	Numeric
5	Tenth Board	Categorical	Tamil Nadu/Kerala/CBSE/Others
6	Higher Secondary Percentage	Continuous	Numeric
7	Higher Secondary Board	Categorical	Tamil Nadu/Kerala/CBSE/Others
8	Undergraduate Course Discipline	Categorical	Arts/Engineering/Science
9	Undergraduate Percentage	Continuous	Numeric
10	Break in Study	Categorical	Yes/No
11	Work Experience	Continuous	Numeric
12	Entrance exam score	Continuous	Numeric
13	Marks obtained in Organizational Behavior course	Continuous	Numeric
14	Marks obtained in Business Environment course	Continuous	Numeric
15	Marks obtained in Managerial Economics course	Continuous	Numeric
16	Marks obtained in Accounting for managers course	Continuous	Numeric
17	Marks obtained in Business Communication course	Continuous	Numeric
18	Marks obtained in Quantitative Techniques course	Continuous	Numeric
19	Marks obtained in Strategic Management course	Continuous	Numeric

4.3 Partition the Data

The data was partitioned into two sets, the training dataset and testing dataset. Eighty percentage of the dataset was used for training the model, and the model was tested using the remaining twenty percentage of the data set.

4.4 Build Models

Logistic regression was used to predict the response variable. Logistic regression is a generalized linear model, where the response variable is a function of the linear combination of all the predictor variables (Lauria and Baron 2011). The categorical predictor variables used in this study include age, gender, community, tenth board, higher secondary board, undergraduate degree and break in study. The continuous predictor variables include tenth percentage, higher secondary percentage, undergraduate percentage, work experience and entrance exam score. The response variable in this study is the academic status which is denoted by pass (50% or more marks) or fail (less than 50% marks) (Palmer 2013; Barker and Sharkey 2012). The academic status of all the six foundation courses was predicted using this model. The prediction of students' marks in capstone course is done using stepwise multiple linear regression. The coefficients derived for each of the predictor variables helped in identifying which of these variables influences the response variable.

4.5 Evaluate the Models

The accuracy of the logistic regression model can be evaluated based on three metrics such as accuracy, specificity and sensitivity which are derived from the confusion matrix. In this study the overall accuracy [(TP + TN)/(TP + TN + FP + FN)] is not considered as a metric for model evaluation. Instead the focus is on sensitivity [TP/(TP + FN)] and specificity [(TN/(TN + FP)], where TP stands for True Positive, TN for True Negative, FP for False Positive and FN for False Negative. The multiple linear regression model developed using stepwise regression was validated through tenfold cross-validation technique. This technique uses ten rounds of cross-validation using multiple cross-validation training and testing sets. The result of this technique estimates the validity of the machine learning model.

Table 2 Frequency analysis

Features	Categories	Frequency	Percentage
Gender	Male	286	55
	Female	236	45
Community	OC	72	14
	BC	272	52
	Others	178	34
Tenth board	Tamil Nadu	367	70
	Kerala	19	4
	CBSE	93	18
	Others	43	8
Higher Secondary Board	Tamil Nadu	411	79
	Kerala	22	4
	CBSE	65	12
	Others	24	5
Undergraduate Degree	Arts	161	31
	Engineering	325	62
	Science	36	7
Break in Study	Yes	201	39
	No	321	61

5 Results and Discussion

5.1 *Descriptive Analytics*

The results of frequency analysis depicted in Table 2 showed almost equal representation of male students (55%) and female students (45%). Descriptive analysis related to the socio-economic status of the students showed that 52% of the students belonged to BC category. Students who have undergone tenth and higher secondary from Tamil Nadu Board represented 70% and 79%, respectively. Students with Engineering as their undergraduate study was 62% which is the highest compared to Arts (31%) and Science (7%). Thirty-nine percentage of the students have a break in their study indicating that they would have taken up a job after their undergraduation.

5.2 *Scatter Plot*

Before deciding on the modeling strategy, it was essential to understand the correlation between response and predictor variables. A scatter plot was used to identify the correlation. The scatter plot between the response variable (marks obtained in strategic management course) and other predictor variables like tenth percentage, higher secondary percentage, undergraduate percentage, work experience, entrance exam scores is shown in Fig. 1. Figure 1 shows a positive relationship between marks obtained in strategic management course and the other predictor variables.

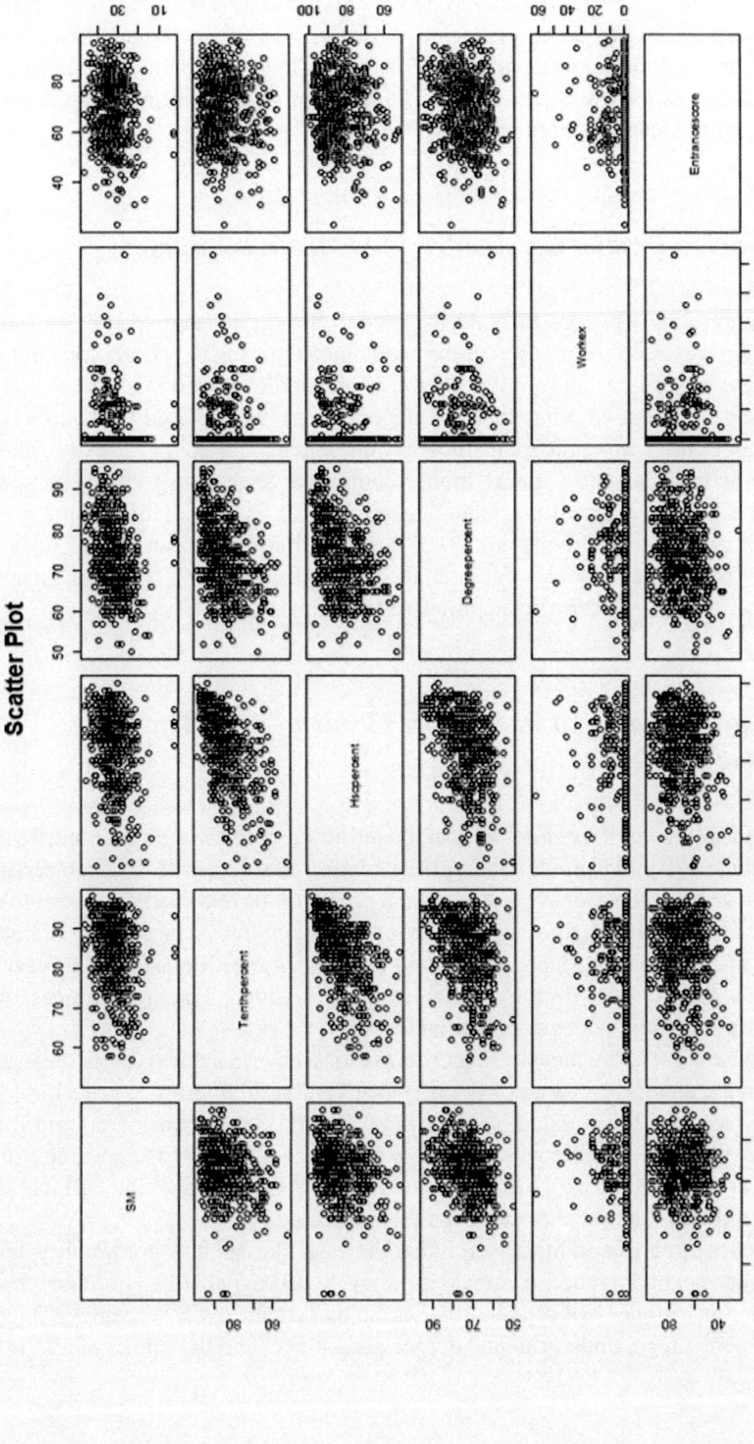

Fig. 1 Scatter plot of variables derived from admission portal

The scatter plot between the response variable (marks obtained in strategic management course) and predictor variables (marks obtained in foundation courses) is shown in Fig. 2. Figure 2 shows a positive relationship between marks obtained in strategic management courses and marks of foundation courses like organizational behavior, business environment, managerial economics, accounting for managers, business communication and quantitative techniques.

5.3 *Model to Predict the Academic Status in a Course*

Logistic regression was used to build the model. Sensitivity and specificity scores of this model was used for predicting the academic status of the foundation courses. In this study, it is important to note that the model so developed should be capable of predicting a student who has failed in the course as "FAIL," and more important that the model *does not* predict a student who actually failed as "PASS." Therefore, specificity as a metric gains more prominence than sensitivity. The specificity of the logistic regression models developed for the foundation courses was 82.54% for organizational behavior, 59.57% for business environment, 92.75% for managerial economics, 60.42% for accounting for managers, 92.47% for business communication and 87.01% for quantitative techniques.

5.4 *Model Building to Predict the Grade of a Student in the Capstone Course*

Multiple linear regression model was deployed to capture the unique contribution of the predictor variables in explaining the variation in the response variable. Since this study had six categorical variables which could not be included directly into the model, the categorical variables were re-coded using dummy variables. For example, since the variable undergraduate degree had three categories, i.e., Arts, Engineering and Science, two $(n - 1)$ dummy variables were included. The same process was applied for all the other categorical variables.

In stepwise regression, the entering criterion for a new variable to enter the model is based on the smallest p value of the partial F test and the removal criterion for a variable is based on the β value. In this study, $\alpha = 0.05$ was considered and if the p value $< \alpha$ then the variable was entered in to the model, and if the p value $> \beta = 0.10$, the variable will be excluded from the model. At each stage, the variable was either entered into the model or removed from the model.

The stepwise regression model excluded the variables such as community, tenth board, higher secondary board, break in study, work experience, entrance exam scores. The final model had retained the variables gender, tenth percentage, higher secondary percentage, undergraduate degree percentage, and the marks of all the six

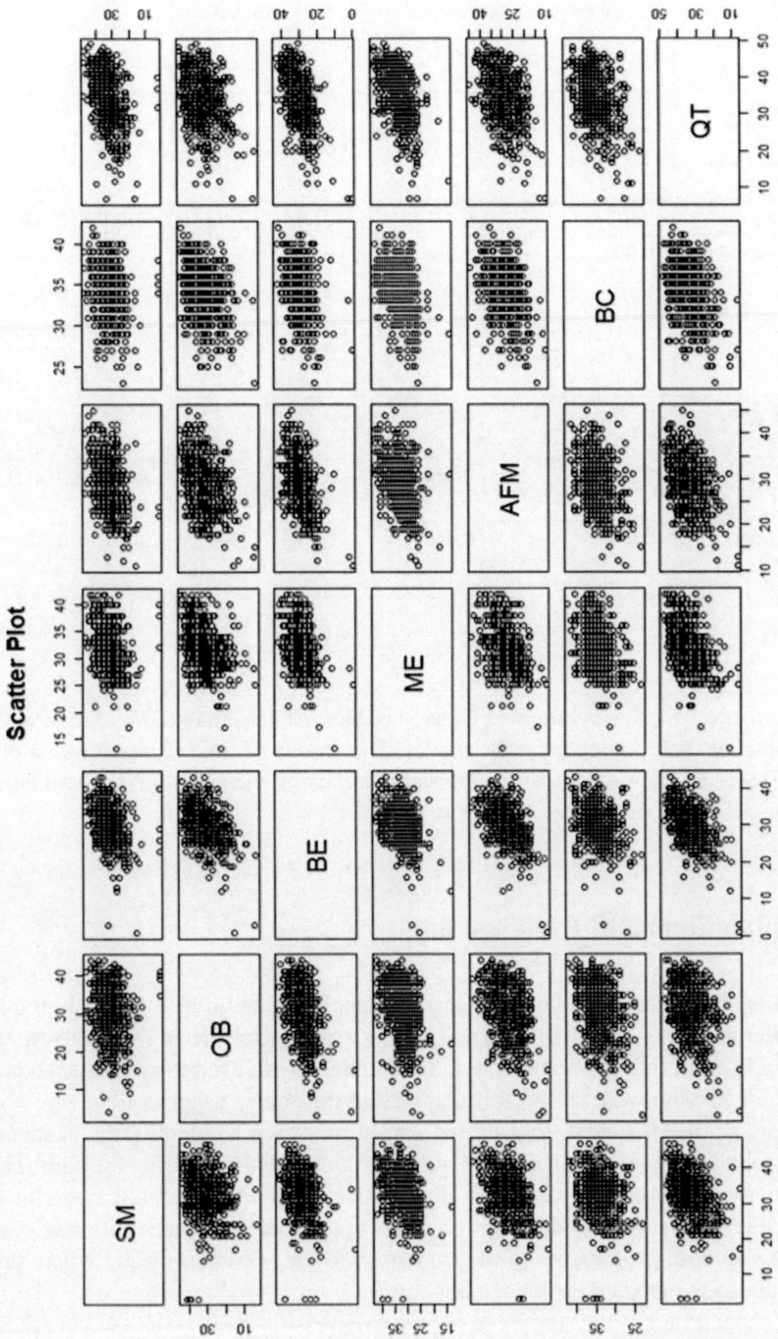

Fig. 2 Scatter plot of variables derived from examination portal

Table 3 Results of stepwise regression model

Variables	Unstandardized coefficients		Standardized coefficients	*t*	Sig
	B	Std. Error	Beta		
(Constant)	11.439	3.148		3.633	0.000
Gender	0.454	0.535	0.036	1.949	0.007
Tenth percent	0.012	0.035	0.017	2.343	0.001
Hsc percent	0.015	0.033	0.023	2.470	0.008
Degree percent	0.070	0.036	0.096	1.938	0.043
Business environment	0.103	0.061	0.087	2.701	0.035
Managerial economics	0.110	0.069	0.076	2.597	0.001
Accounting for Managers	0.229	0.053	0.211	4.327	0.000
Business communication	0.172	0.044	0.198	3.897	0.000
Organizational behavior	0.084	0.039	0.096	2.142	0.033
Quantitative techniques	0.122	0.090	0.062	2.357	0.005

foundation courses. The *p* values of these variables are less than 0.05 indicating the significance of these variables in the model. The results of final regression model is given in Table 3. The model was further validated using the tenfold cross-validation technique. The root mean square error is 5.6.

6 Implications and Conclusion

The results of this study indicate that learning analytics could be effectively implemented in enhancing the quality of teaching–learning experience (Macfadyen and Dawson 2012). In this paper, two predictive models were used to predict academic risk of students who were not performing well in the course as an early intervention mechanism. In the first part, logistic regression was used to identify the academic status of foundation courses in the first semester. Data obtained during the admission process is used as input for model building. Since an MBA program is open for all streams of undergraduate studies, it is essential to have an early intervention in order to ensure a smooth progression of the students into the second semester where they are introduced to advanced management courses.

In the second part of the study, the stepwise regression model was used to predict the marks of the students in capstone course. The results showed that as the students' progress into second semester courses, the tenth and higher secondary board become irrelevant. Performance in the first semester courses greatly influences the results of the second semester. The student who scored well in the first semester also scored well in the second semester. Therefore, this early intervention would help enhance student performance, thereby preparing him to face forthcoming semesters more confidently. This understanding further helps students to select courses in which they can perform better.

Model deployment would help build a transparent system by which both the stakeholders, faculty and student would get insights about the students' progress. This study could be further extended to all courses in the forthcoming semesters. This would gradually evolve into a learning analytics system which can be inbuilt in to the curriculum. Further, this model could be extended to predict the probability of the students succeeding in placement. Deployment of the models developed in this study would go a long way in not only enhancing students' performance but also more fruitful faculty engagement. Embedding analytics in the education system would transform the education landscape to greater heights.

References

Agudo-Peregrina, Á. F., Iglesias-Pradas, S., Conde-González, M. Á., & Hernández-García, Á. (2014). Can we predict success from log data in VLEs? Classification of interactions for learning analytics and their relation with performance in VLE-supported F2F and online learning. *Computers in Human Behavior, 31,* 542–550.

Arnold, K. E., & Pistilli, M. D. (2012). Course signals at Purdue: Using learning analytics to increase student success. In *Proceedings of the 2nd international conference on learning analytics and knowledge* (pp. 267–270). ACM.

Baker, R., & Siemens, G. (2014). Educational data mining and learning analytics. In R. K. Sawyer (Ed.), *Cambridge handbook pf the learning sciences.* Cambridge, UK: Cambridge University Press.

Baker, R. S., & Inventado, P. S. (2014). Educational data mining and learning analytics. In *Learning analytics* (pp. 61–75). Springer, New York, NY.

Baradwaj, B. K., & Pal, S. (2011). Mining educational data to analyze students' performance. *International Journal of Advanced Computer Science and Applications, 2*(6), 63–69.

Barber, R., & Sharkey, M. (2012), Course correction: Using analytics to predict course success. In *Proceedings of the 2nd international conference on learning analytics and knowledge* (pp. 259–262). ACM.

Black, A. E., & Deci, E. L. (2000). The effects of instructors' autonomy support and students' autonomous motivation on learning organic chemistry: A self-determination theory perspective. *Science Education, 84*(6), 740–756.

Campbell, J. P. (2007), *Utilizing student data within the course management system to determine undergraduate student academic success: An exploratory study*. Purdue University.

Chung, S. H., Schwager, P. H., & Turner, D. E. (2002). An empirical study of students' computer self-efficacy: Differences among four academic disciplines at a large university. *Journal of Computer Information Systems, 42*(4), 1–6.

Elbadrawy, A., Polyzou, A., Ren, Z., Sweeney, M., Karypis, G., & Rangwala, H. (2016). Predicting student performance using personalized analytics. *Computer, 49*(4), 61–69.

Ferguson, R. (2012). Learning analytics: Drivers, developments and challenges. *International Journal of Technology Enhanced Learning, 4*(5/6), 304–317.

Gašević, D., Dawson, S., & Rogers, T. (2016). Learning analytics should not promote one size fits all: The effects of instructional conditions in predicting academic success. *The Internet and Higher Education, 28,* 68–84.

Han, J., & Kamber, M. (2006). *Data mining: Concepts and techniques* (2nd ed.). Boston, MA: Elsevier.

Lauría, E. J., & Baron, J. (2011). Mining Sakai to measure student performance: Opportunities and challenges in academic analytics. *Download at:* https://eccmarist.edu/conf2011/materials.

Laurie, P. D., & Timothy, E. (2005). Using data mining as a strategy for assessing asynchronous discussion forums. *Computer Education, 45*(1), 141–160.

Lykourentzou, I., Giannoukos, I., Mpardis, G., Nikolopoulos, V., & Loumos, V. (2009). Early and dynamic student achievement prediction in e-learning courses using neural networks. *Journal of the American Society for Information Science and Technology, 60*(2), 372–380.

Macfadyen, L. P., & Dawson, S. (2012), Numbers are not enough. Why e-learning analytics failed to inform an institutional strategic plan. *Journal of Educational Technology & Society, 15*(3).

Palmer, S. (2013). Modelling engineering student academic performance using academic analytics. *International Journal of Engineering Education, 29*(1), 132–138.

Romero, C., López, M. I., Luna, J. M., & Ventura, S. (2013). Predicting students' final performance from participation in on-line discussion forums. *Computers and Education, 68,* 458–472.

Siemens, G., Gasevic, D., Haythornthwaite, C., Dawson, S. P., Shum, S., Ferguson, R, & Baker, R. (2011). Open learning analytics: An integrated & modularized platform.

Talavera, L., & Gaudioso, E. (2004). Mining student data to characterize similar behavior groups in unstructured collaboration spaces. In *Workshop on artificial intelligence in CSCL. 16th European conference on artificial intelligence* (pp. 17–23).

Whitmire, E. (2002). Disciplinary differences and undergraduates' information-seeking behavior. *Journal of the American Society for Information Science and Technology, 53*(8), 631–638.

An Optimal Response-Adaptive Design for Multi-treatment Clinical Trials with Circular Responses

Taranga Mukherjee, Rahul Bhattacharya, and Atanu Biswas

1 Introduction

In clinical trials, equal allocation for assignment of subjects to competing treatment arms has long been advocated by the medical practitioners to reflect the view of equipoise at the outset of the trial. But continuing the view of equipoise Rosenberger and Lachin (2002) balances the allocation among the treatments without making any distinction between the superior and inferior treatments. Such lack of distinction among treatments under consideration is naturally questionable from ethical point of view and suggests continuous monitoring coupled with dynamic allocation. A dynamic allocation procedure allows the experimenter to evaluate the treatments at intermediate stages of the trial and skews the allocation in favour of the treatment doing better of the trial based on the available data. If the available allocation and response data are used for the allocation of every incoming subject, the allocation is termed a response-adaptive allocation.

Most of the response-adaptive designs, available in the literature, are developed for two treatment trials and only a few are available for multiple treatments. Further, almost all the available response-adaptive designs are either for binary, categorical or conventional continuous (often termed "linear") treatment responses. But angular responses are the natural outcomes in the context of several biomedical studies (e.g. in orthopedics, ophthalmology, sports medicine). The usual (i.e. linear) contin-

T. Mukherjee (✉) · R. Bhattacharya
Department of Statistics, University of Calcutta, Kolkata, India
e-mail: tm.custat@gmail.com

R. Bhattacharya
e-mail: rahul_bhattya@yahoo.com

A. Biswas
Applied Statistics Unit, Indian Statistical Institute, Kolkata, India
e-mail: atanu@isical.ac.in

A. K. Laha (ed.), *Applied Advanced Analytics*, Springer Proceedings in Business and Economics, https://doi.org/10.1007/978-981-33-6656-5_13

uous probability distributions are, therefore, inappropriate to capture periodicity in a bounded domain Fisher (1993). Naturally, applying an allocation design for linear continuous responses for circular response trials is not only inappropriate but may lead to fallacious results. Despite several occurrences of circular data in clinical trials, application of response-adaptive allocation designs in trials involving circular responses has attracted less attention Atkinson and Biswas (2014). Further, designing a clinical trial should not only focus on the ethical requirements (i.e. assigning higher number of subjects to the eventually better treatment) but efficiency issues (e.g. making precise inference on treatment efficacy) are also equally important. Considering both the ethical and efficiency requirements within the same framework, Biswas et al. (2015) and Biswas et al. (2017) developed a two-treatment allocation design for circular response clinical trials, which is one of the earliest contributions in this class of allocation designs and is commonly known as an optimal response-adaptive allocation. But clinical trials may involve multiple treatments, and defining ethical and efficiency concerns in the presence of several treatments is met with different challenges. In the current work, we define appropriate ethics and efficiency measures assuming multiple treatments and derive an optimal allocation design by weighing such requirements in a sensible way. In Sect. 2, we develop the ethics and efficiency requirements for circular response models and considering a constrained optimization problem, derive the optimal target proportion. We provide the response-adaptive randomization to target the optimal proportion in practice along with related large sample results in Sect. 3. Small sample performance of the proposed design is investigated and compared with the "gold standard" equal allocation in detail in Sect. 4. In Sect. 5, we redesign a real clinical trial with circular outcome adopting the proposed allocation design. Some related and relevant issues are finally discussed in Sect. 6.

2 The Proposed Allocation Design

Consider a clinical trial with t (>2) competing treatments, where the patient outcome is circular in nature. Unlike linear responses , circular responses cannot be compared directly and hence identifying a "better" patient response requires further consideration. In fact, circular responses are periodic in nature and hence in circular set-up, the responses 20° and 340° are identical in effect. Consequently, fallacious conclusions may be reached if such responses are analysed using existing methods. To avoid such impediment, the comparison among circular treatment responses is made with respect to a preferred direction, which is treated as a reference point. In general, a preferred direction should be chosen by practitioners as per the requirement of the study. A preferred direction can be chosen in multiple ways. For example, in medical studies related to shoulder movement, it is usually seen that a perfect shoulder allows 90° of internal rotation (Jain et al. 2013), and the preferred direction should be taken as 90° in that context. However, preferred direction can also be data driven.

Once a preferred direction is set, intuitively, a treatment is promising if it produces responses near the preferred direction. Therefore, if μ_0 is set as the preferred direction in a clinical trial, the quality of a response is determined by an appropriate distance from the preferred direction. Due to the periodic nature of circular responses, the linear deviation of the responses from the preferred direction yields little or no sense. We, therefore, use a circular distance measure Jammalamadaka and SenGupta (2001) defined by smaller of the two arc lengths between the preferred angle and the response angle along the circumference of an unit circle. Analytically, the circular distance between an arbitrary circular response ψ and preferred direction μ_0 can be expressed as $d(\psi, \mu_0) = \min(\psi - \mu_0, 2\pi - \overline{\psi - \mu_0})$ (see, Jammalamadaka and SenGupta (2001), for example). The distance d is a linear quantity having no periodicity and hence can be ordered conventionally. However, we have kept the preferred direction at 0°, throughout the work for the sake of brevity.

Since the aim of the current work is to develop allocation designs considering both ethics and efficiency, we need appropriate measures of both. For ethics, we introduce a clinically meaningful threshold "c", the distance above which is regarded as a treatment failure. Specifically, an observed response ψ is regarded beneficial if $d(\psi, 0) \leq c$. Therefore, if we consider a hypothetical non-randomized clinical trial with t treatments and n_k assignments to treatment k, the expected total number of benefited subjects is

$$\sum_{k=1}^{t} n_k P\{d(Y_k, 0) \leq c\}$$

where Y_k represents the responses to treatment k. Naturally, a higher value of the above or equivalently a lower value of $H(n_1, n_2, \ldots n_t) = \sum_{k=1}^{t} n_k P\{d(Y_k, 0) > c\}$ is desirable from an ethical perspective.

However, to measure the efficiency, we consider A optimality ((Silvey, 1980)) based on the large sample dispersion matrix of $(d(\tilde{\mu_1}, 0), d(\tilde{\mu_2}, 0), \ldots, d(\tilde{\mu_t}, 0))^T$, where $\tilde{\mu}_k$ is an estimator of μ_k under the non-randomized allocation. The large sample dispersion matrix takes the form $Diag(\frac{\sigma_1^2}{n_1}, \frac{\sigma_2^2}{n_2}, ..., \frac{\sigma_t^2}{n_t})$, with $\frac{\sigma_k^2}{n_k}$ as the large sample variance of $d(\tilde{\mu_k}, 0)$. Then, A optimality criterion dictates to use $\sum_{k=1}^{t} \frac{\sigma_k^2}{n_k}$ as an efficiency measure, where a lower value indicates higher precision of estimators. Thus, we suggest to obtain the optimal proportion $\rho_k = \left(\frac{n_k}{\sum_{j=1}^{t} n_j}\right)_{opt}$ to treatment k by solving the constrained optimization problem:

$$\text{Minimize } \sum_{k=1}^{t} \frac{\sigma_k^2}{n_k},$$
$$\text{subject to } \sum_{k=1}^{t} n_k P(d(Y_k, 0) > c) < h$$

for some $h > 0$. Application of standard optimization techniques ((Bazaraa et al., 2006)) expresses the optimal proportion as

$$\rho_k = \frac{\frac{\sigma_k}{\sqrt{\gamma_k}}}{\sum_{k=1}^{t} \frac{\sigma_k}{\sqrt{\gamma_k}}}, k = 1, 2, \ldots t.$$

where $\gamma_k = P\{d(Y_k, 0) > c\}$.

3 Implementation of the Allocation Design in Practice

The optimal target allocation function ρ_k is naturally a function of the parameters of the response distribution, and hence implementation requires knowledge of such unknown quantities. But these unknown parameters (say, θ) are never known in advance. Hence, we suggest to design the trial adaptively; that is, we suggest to use the currently available response and allocation data to estimate θ. In adaptive allocation, initially n_0 subjects are allocated to each of the t treatment arms, then responses from tn_0 subjects are obtained and based on that information the allocation probability for $(tn_0 + 1)$th subject is calculated. Naturally, this initial allocation n_0 is kept lower to assign more subjects adaptively.

Suppose $\delta_{k,i}$ is the treatment indicator taking the values 1 or 0 accordingly as the ith subject is assigned treatment k or not, and $\mathcal{F}_i$ indicates the information contained in the allocation-and-response data obtained up to and including the ith subject. Then, the $(i + 1)$th subject is assigned to treatment k with probability

$$P(\delta_{k,i+1}|\mathcal{F}_i) = \rho_k(\widehat{\boldsymbol{\theta}}^{(i)}),$$

where $\rho_k(\widehat{\boldsymbol{\theta}}^{(i)})$ is a strongly consistent estimator of ρ_k based on the available data up to and including the ith subject. In practice, we use sequentially updated maximum likelihood estimators and plug it into the allocation function at every stage to calculate the allocation probabilities.

Since for any allocation design, primary concern is ethics, we study the behaviour of the observed proportion of allocation to different treatments. If we denote the number of allocations by the proposed design to treatment k out of n assignments by $N_{kn} = \sum_{i=1}^{n} \delta_{k,i}$, the observed allocation proportion to treatment k is simply $\frac{N_{kn}}{n}$. Then under certain widely satisfied restrictions Hu et al. (2004) on the response distribution and continuity of $\rho_k(\theta_1, \theta_2, \ldots, \theta_k)$ in each of its arguments for every $k = 1, 2, \ldots, t$, we have the following result.

Result: As $n \to \infty$

$$\frac{N_{kn}}{n} \to \rho_k(\boldsymbol{\theta})$$

almost surely for each k = 1,2, …t

4 Performance Evaluation

4.1 Performance Measures

Performance of any allocation design needs to be assessed in the light of both ethics and optimality. An allocation function exhibits strong ethical perspective if it allocates higher number of patients to the better performing treatment arm. In this context, the expected allocation proportions (EAPs), defined by $E(\frac{N_{kn}}{n})$, $k = 1, 2, \ldots, t$, can be regarded as a measure of ethics, where the higher the value of EAP for better performing treatment arm is the indicator for ethical impact of the allocation design. Again to measure efficiency, we use the power of a relevant test of equality of treatment effects. But the concerned test is not a simple adaptation of the usual test of homogeneity for linear responses. In fact, for circular responses if μ_k is the mean direction associated with the kth treatment, then treatments j and k are equally effective if $d(\mu_k, 0) = d(\mu_j, 0)$ or equivalently if $\mu_k = \mu_j(\mod 2\pi)$ or $\mu_k = 2\pi - \mu_j(\mod 2\pi)$. However, the distance functions are linear in nature, and hence as an alternative, we consider testing the null
$H_0 : d(\mu_1, 0) = d(\mu_2, 0) = \cdots d(\mu_t, 0)$ against the alternative H_1 : at least one inequality in H_0.

Assuming treatment 1 as experimental and others as existing, we define the contrast-based homogeneity test statistic

$$T_n = (\mathbf{H}\hat{\mathbf{d}})^T \left[\mathbf{H}\hat{\Sigma}_{\hat{\mathbf{d}}}\mathbf{H}^T\right]^{-1} (\mathbf{H}\hat{\mathbf{d}}),$$

where

$$\hat{\mathbf{d}}^{t\times 1} = \begin{pmatrix} d(\hat{\mu}_1, 0) \\ d(\hat{\mu}_2, 0) \\ \vdots \\ d(\hat{\mu}_t, 0) \end{pmatrix},$$

$$\mathbf{H}^{t-1\times t} = \begin{bmatrix} 1 & -1 & 0 & \cdots & 0 \\ 1 & 0 & -1 & \cdots & 0 \\ \vdots & \vdots & & \ddots & \vdots \\ 1 & 0 & 0 & \cdots & -1 \end{bmatrix},$$

$\hat{\Sigma}_{\hat{d}}$ is the estimated dispersion matrix of $\hat{d}^{t\times 1}$, and $\hat{\mu}_k$ is a strongly consistent estimator of μ_k based on n observations, generated through the proposed adaptive allocation design. Naturally, larger value of T_n indicates departure from the null hypothesis and hence a right tailed test based on T_n is appropriate to test H_0 against H_1.

4.2 Simulation Studies

In order to evaluate performance of the proposed optimal allocation design, we consider three treatments, namely 1, 2 and 3, and keep treatment 1 as the superior one (i.e. having minimum circular distance from preferred direction 0°) followed by treatments 2 and 3. Specifically, we assume that the response distribution for treatment j is von Mises with mean direction μ_j and concentration κ_j, $j = 1, 2, 3$. Naturally, μ_1 is kept closest to 0°. However, the concentration parameters κ_1, κ_2 and κ_3 are varied accordingly. Three sets of concentration parameters are considered separately. First, all treatment arms are assumed to have equal concentrations, then higher concentration is assigned to the superior treatment arms and finally higher concentration is assigned to inferior treatment arm. Considering different configurations of (μ_1, μ_2, μ_3) and $(\kappa_1, \kappa_2, \kappa_3)$, we conduct a simulation study with 25,000 iterations. The simulation is carried out for both the choices $n = 60$ and $n = 240$. For $n = 240$, the initial equal allocation n_0 is kept as 10, and for $n = 60$ the same is kept at 3. However for evaluation, the threshold value is fixed at $c = 30°$

Since the power of the concerned test under equal allocation is often considered as a benchmark, the power under the proposed optimal allocation is compared with that of equal allocation, where each treatment is assigned with equal probability. The details of expected allocation proportion (EAP) and power are reported in Tables 1 and 2. The performance figures in Tables 1 and 2 indicate that the allocation function successfully assigns a larger number of subjects to the superior treatment arm keeping the power almost as good as that of equal allocation. Also, the corresponding standard

Table 1 Expected allocation proportion and power for $n = 60$

$\mu_1, \mu_2, \mu_3, \kappa_1, \kappa_2, \kappa_3$	EAP(SD)			Power	
	1	2	3	Proposed	Equal
(5, 5, 5, 2.0, 2.0, 2.0)	0.333(0.09)	0.333(0.09)	0.333(0.09)	0.050	0.050
(5, 10, 15, 2.0, 2.0, 2.0)	0.345(0.09)	0.331(0.07)	0.322(0.08)	0.141	0.118
(5, 15, 25, 2.0, 2.0, 2.0)	0.348(0.07)	0.332 (0.07)	0.318(0.08)	0.324	0.814
(5, 30, 45, 2.0, 2.0, 2.0)	0.366 (0.07)	0.323(0.08)	0.310 (0.09)	0.833	0.810
(5, 45, 60, 2.0, 2.0, 2.0)	0.370(0.07)	0.318(0.07)	0.311 (0.08)	0.975	0.097
(5, 75, 90, 2.0, 2.0, 2.0)	0.377 (0.07)	0.311 (0.07)	0.313(0.07)	1.000	1.000
(5, 5, 5, 1.0, 2.0, 2.0)	0.434(0.07)	0.288 (0.07)	0.280(0.06)	0.05	0.05
(5, 10, 15, 1.0, 2.0, 2.0)	0.439(0.08)	0.282 (0.07)	0.277 (0.08)	0.070	0.090
(5, 30, 45, 1.0, 2.0, 2.0)	0.462 (0.08)	0.275 (0.08)	0.261(0.08)	0.420	0.248
(5, 45, 60, 1.0, 2.0, 2.0)	0.472 (0.07)	0.267(0.07)	0.260 (0.07)	0.778	0.664
(5, 75, 90, 1.0, 2.0, 2.0)	0.474 (0.07)	0.262(0.07)	0.263 (0.07)	0.993	0.969
(5, 5, 5, 2.0, 2.0, 1.0)	0.292(0.07)	0.280 (0.07)	0.424(0.07)	0.050	0.050
(5, 15, 25, 2.0, 2.0, 1.0)	0.311(0.08)	0.276(0.08)	0.411(0.07)	0.203	0.112
(5, 30, 45, 2.0, 2.0, 1.0)	0.311(0.07)	0.276(0.06)	0.414(0.07)	0.530	0.269
(5, 45, 60, 2.0, 2.0, 1.0)	0.317(0.08)	0.267(0.06)	0.411(0.07)	0.800	0.634
(5, 75, 90, 2.0, 2.0, 1.0)	0.317(0.07)	0.268(0.07)	0.419(0.07)	0.985	0.991

Table 2 Expected allocation proportion and power for $n = 240$

$\mu_1, \mu_2, \mu_3, \kappa_1, \kappa_2, \kappa_3$	EAP(SD)			Power	
	1	2	3	Proposed	Equal
(5, 5, 5, 2.0, 2.0, 2.0)	0.333(0.04)	0.333(0.05)	0.333(0.04)	0.050	0.050
(5, 10, 15, 2.0, 2.0, 2.0)	0.342(0.04)	0.334(0.04)	0.323(0.04)	0.339	0.355
(5, 15, 25, 2.0, 2.0, 2.0)	0.352(0.03)	0.331(0.04)	0.316(0.04)	0.816	0.855
(5, 30, 45, 2.0, 2.0, 2.0)	0.365(0.05)	0.323(0.04)	0.311(0.04)	1.000	1.000
(5, 5, 5, 1.0, 2.0, 2.0)	0.458(0.04)	v.272(0.05)	0.278(0.05)	0.050	0.050
(5, 10, 15, 1.0, 2.0, 2.0)	0.468(0.05)	0.269(0.05)	0.262(0.04)	0.130	0.060
(5, 30, 45, 1.0, 2.0, 2.0)	0.491(0.04)	0.258(0.05)	0.249(0.04)	0.987	0.944
(5, 45, 60, 1.0, 2.0, 2.0)	499(0.04)	0.252(0.05)	0.247(0.04)	1.000	1.000
(5, 5, 5, 2.0, 2.0, 1.0)	0.270(0.04)	0.266(0.04)	0.458(0.07)	0.050	0.050
(5, 15, 25, 2.0, 2.0, 1.0)	0.283(0.04)	0.268(0.05)	0.447(0.05)	0.597	0.510
(5, 30, 45, 2.0, 2.0, 1.0)	0.294(0.04)	0.260(0.05)	0.445(0.04)	0.983	0.982
(5, 45, 60, 2.0, 2.0, 1.0)	0.299(0.05)	0.255(0.04)	0.445(0.05)	1.000	1.000

deviations, measuring the allocation fluctuation (reported in the parenthesis in the tables), remain significantly lower irrespective of the chosen sample sizes. All these facts make the proposed optimal allocation rule a competent one.

We further have studied the performance of the proposed allocation design considering four treatments based on limiting allocation proportion (LAP) by varying the threshold value c. LAP essentially indicates the theoretical limiting proportion of number of subjects allocated to a certain treatment arm to the total number of subjects available. The mean direction parameters for treatments 1, 2 and 3 are kept at 5°, 15° and 25°, respectively, and mean direction for treatment 4; i.e. μ_4 is varied from 25° to 160°. Thus, treatments 1, 2 3 and 4 can be regarded as ordered from superior to inferior. Naturally for a sensible allocation design limiting allocation proportion to treatment 1 should increase as μ_4 drifts away from 25°. In Fig. 1, we plot LAP to treatment 1 (i.e. the superior treatment) for varying μ_4 and various choices of $(\kappa_1, \kappa_2, \kappa_3, \kappa_4)$. The plot in Fig. 1 is found to be in agreement with the anticipated behaviour of LAP across various choices of c and $(\kappa_1, \kappa_2, \kappa_3, \kappa_4)$

5 Redesigning a Real Clinical Trial: SICS Trial

Now to evaluate the proposed procedure from a real clinical perspective, we consider a real trial on small incision cataract surgery (Bakshi 2010). We take into account three competing treatments, namely snare technique (see Basti 1993), irrigating vectis technique (see Masket 2004) and torsional phacoemulsification (see Mackool and Brint 2004) based on 19, 18 and 16 observations, respectively. Responses corresponding to each treatment are circular in nature, and hence the trial is appropriate to

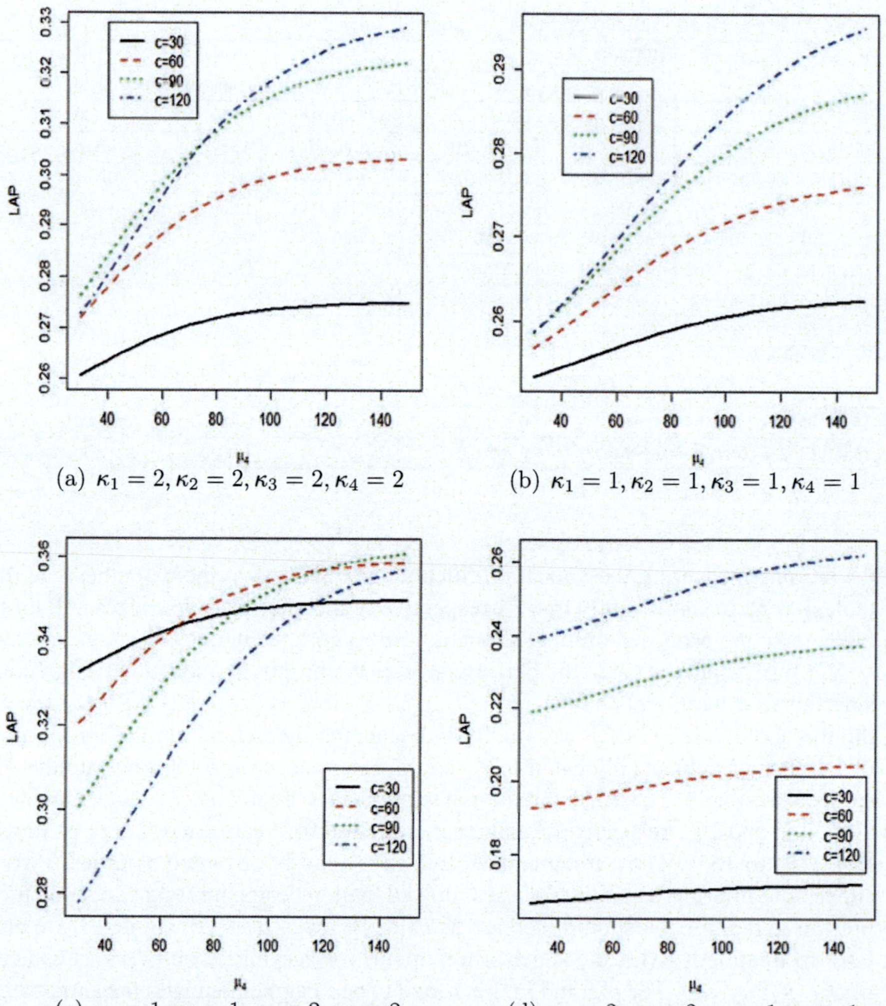

(a) $\kappa_1 = 2, \kappa_2 = 2, \kappa_3 = 2, \kappa_4 = 2$

(b) $\kappa_1 = 1, \kappa_2 = 1, \kappa_3 = 1, \kappa_4 = 1$

(c) $\kappa_1 = 1, \kappa_2 = 1, \kappa_3 = 2, \kappa_4 = 2$

(d) $\kappa_1 = 2, \kappa_2 = 2, \kappa_3 = 1, \kappa_4 = 1$

Fig. 1 Limiting allocation proportion von Mises response for four treatments

judge the performance of the proposed allocation. The responses obtained from these three types of surgical interventions, namely snare, irrigating vectis and torsional phacoemulsification techniques, are assumed to follow von Mises with parameters (μ_s, κ_s), (μ_v, κ_v) and (μ_t, κ_t), respectively, and rationale behind such assumption is verified by Watson's goodness of fit test Mardia and Jupp (2004). In the light of this three independent competing treatments, the proposed allocation design is redesigned with the following parameter choices, estimated from the available data points.

Table 3 Allocation to different treatment arms

Treatment	EAP(SD)	
	Proposed	Actual
Snare	0.250(0.07)	0.358
Irrigating vectis	0.201(0.08)	0.339
Torsional phacoemulsification	0.548(0.08)	0.301

For the snare technique, parameters are estimated as $\hat{\mu}_s = 20.67°$, $\hat{\kappa}_s = 1.59$; for irrigating vectis, these estimates are $\hat{\mu}_v = 52.71°$, $\hat{\kappa}_v = 1.27$; for torsional phacoemulsification, the estimates of the parameters are $\hat{\mu}_t = 2.29°$, $\hat{\kappa}_t = 4.99$, respectively. As far as distance from preferred direction is concerned, torsional phacoemulsification appears to be much better than its competitors followed by snare's technique. In addition, torsional phacoemulsification has significantly higher concentration over others. Thus, the treatment clearly emerges as the best one. From Tables 3 and 4, we find that the proposed optimal allocation design produces about 23% higher EAP to the superior treatment torsional phacoemulsification and reduced the EAP for the other treatments as compared to the original allocation. This naturally shows the ethical impact of the proposed optimal response-adaptive allocation and hence makes the proposed allocation desirable in real clinical trial.

6 Concluding Remarks

The current work develops an optimal treatment allocation design for multiple arms by minimizing total number of treatment failures subject to fixed precision. Although essence of the proposed design is based on ethical point of view, the optimality of inference of treatment effect detection is not compromised. In fact, it is well competing with equal allocation design. However, no covariate effect is studied here, which is left for future consideration.

Acknowledgements The authors of this paper would like to thank the anonymous referees for their valuable comments towards the betterment of the current work.

References

Atkinson, A. C., & Biswas, A. (2014). *Randomised response-adaptive designs in clinical trials*. Boca Raton: CRC Press.

Bakshi, P. (2010). *Evaluation of various surgical techniques in Brunescent cataracts* (Unpublished thesis). Disha Eye Hospital, India.

Basti, S., Vasavada, A. R., Thomas, R., & Padhmanabhan, P. (1993). Extracapsular cataract extraction: Surgical techniques. *Indian Journal Ophthalmology*, *41*, 195–210.

Bazaraa, M., Sherali, H., & Shetty, C. M. (2006). *Nonlinear programming: Theory and algorithms* (3rd edn.). Chichester, London: Wiley.

Biswas, A., Bhattacharya, R., Mukherjee, T. (2017). An adaptive allocation design for circular treatment outcome. *Journal of Statistical Theory and Practice*. https://doi.org/10.1080/15598608.2017.1307147.

Biswas, A., & Coad, D. S. (2005). A general multi-treatment adaptive design for multivariate responses. *Sequential Analysis*, *24*, 139–158.

Biswas, A., Dutta, S., Laha, A. K., & Bakshi, P. K. (2015). Response-adaptive allocation for circular data. *Journal of Biopharmaceutical Statistics*, *25*, 830–842.

Fisher, N. I. (1993). *Statistical analysis of circular data*. Cambridge: Cambridge University Press.

Hu, F., & Zhang, L. X. (2004). Asymptotic properties of doubly adaptive biased coin design for multi-treatment clinical trials. *Annals of Statistics*, *32*, 268–301.

Jain, N. B., Wilcox, III, R. B., & I. I. I., Katz, J. N., & Higgins, L. D., (2013). Clinical examination of the rotator cuff. *PM& R*, *5*, 45–56.

Jammalamadaka, S. R., & SenGupta, A. (2001). *Topics in circular statistics*. Singapore: World Scientific.

Mackool, R. J., & Brint, S. F. (2004). AquaLase: A new technology for cataract extraction. *Current Opinion Ophthalmology*, *15*, 40–43.

Mardia, K. V., & Jupp, P. E. (2004). *Directional statistics*. Chichester, London: Wiley.

Masket, S. (2004). The beginning of modern cataract surgery. The evaluation of small incision cataract surgery—A short history of ophthalmologists in progress. *Cataract and Refractory Surgeries Today*, 77–80.

Rosenberger, W. F., & Lachin, J. L. (2002). *Randomisation in clinical trials: Theory and practice*. New York: Wiley.

Silvey, S. (1980). *Optimal designs: An introduction to the theory for parameter estimation*. Springer Texts.

Stochastic Comparisons of Systems with Heterogeneous Log-Logistic Components

Shyamal Ghosh, Priyanka Majumder, and Murari Mitra

1 Introduction

The log-logistic distribution, henceforth referred to as LLD, is a well-known life distribution that finds widespread application in different fields such as survival analysis, hydrology, economics, and networking.

It has been used in regression models for survival data (see Bennet 1983) and also as a parametric model for events whose failure rate increases initially and decreases later, for example, the mortality rate from cancer following diagnosis or treatment. Its application can also be seen in the field of hydrology for modeling precipitation and stream flow rates. For example, to analyze Canadian precipitation data Shoukri et al. (1988) showed that LLD is a suitable choice whereas Fahim and Smail (2006) used LLD for modeling stream flow rates. The LLD is also known as Fisk distribution in the field of economics where it has been utilized to describe the distribution of wealth or income (see Fisk 1961). In the field of computer science and networking, LLD has been used as a more accurate probabilistic model (see Gago-Benítez et al. 2013 for details).

The LLD is very similar in shape to the log-normal distribution but has the added advantage of being mathematically more tractable because of its closed form dis-

S. Ghosh (✉)
Department of Mathematical Statistics and Actuarial Science, University of the Free State, Bloemfontein, South Africa
e-mail: shyamalmath2012@gmail.com

P. Majumder
Department of Mathematics, Indian Institute of Technology Bombay, Mumbai, India
e-mail: priyankamjmdr@gmail.com

M. Mitra
Department of Mathematics, Indian Institute of Engineering Science and Technology, Howrah, India
e-mail: murarimitra@yahoo.com

A. K. Laha (ed.), *Applied Advanced Analytics*, Springer Proceedings in Business and Economics, https://doi.org/10.1007/978-981-33-6656-5_14

tribution function and its quite flexible hazard rate function. It is a good alternative to the Weibull, whose hazard rate function is either increasing or decreasing, i.e., monotonic, depending on the value of its shape parameter. As such, the use of the Weibull distribution may be inappropriate where the course of the disease is such that mortality reaches a peak after some finite period and then slowly declines. Additionally, the LLD is also connected to extreme value distributions. As showed by Lawless (1986), the Weibull distribution has paramount importance in reliability theory as it is the only distribution that belongs to two families of extreme value distributions, each of which has essential qualities for the study of proportional hazard and accelerated failure times. Thus, the LLD possesses the nice characteristic of being a representative of both these families.

A random variable (r.v.) X is said to have the LLD with shape parameter α and scale parameter γ, written as $LLD(\alpha, \gamma)$, if its *probability density function* (pdf) is given by

$$f(x;\alpha,\gamma) = \frac{\alpha\gamma(\gamma x)^{\alpha-1}}{(1+(\gamma x)^{\alpha})^2}\ ,\ x \geq 0,\ \ (\alpha > 0, \gamma > 0). \tag{1.1}$$

Just as one gets the log-normal and log-Pearson distributions from normal and Pearson distribution, LLD is obtained by taking the logarithmic transformation of the logistic distribution. The LLD is also a special case of the 'kappa distributions' introduced by Mielke and Johnson (1973). Another interesting fact is that LLD can also be obtained from the ratio of two independent Stacy's generalized gamma variables (see Malik 1967; Block and Rao 1973). Even though different properties of this distribution have been explored intensely by many researchers, the stochastic comparisons of their extreme order statistics have not been studied so far. This is the primary motivation behind the present work.

But first, a few words about order statistics which occupy a place of remarkable importance in both theory and practice. It play a vital role in many areas including reliability theory, economics, management science, operations research, insurance, hydrology, etc., and have received a lot of attention in the literature during the last several decades [(see, e.g., the two encyclopedic volumes by Balakrishnan and Rao (1998a, b)]. Let $X_{1:n} \leq \cdots \leq X_{n:n}$ represent the order statistics corresponding to the n independent random variables (r.v.'s) $X_1, \ldots, X_n$.

It is a well-known fact that the kth order statistic $X_{k:n}$ represents the lifetime of a $(n-k+1)$-out-of-n system which happens to be a suitable structure for redundancy that has been studied by many researchers. Series and parallel systems, which are the building blocks of many complex coherent systems, are particular cases of a k-out-of-n system. A series system can be regarded as a n-out-of-n system, while a parallel system is a 1-out-of-n system. In the past two decades, a large volume of work has been carried out to compare the lifetimes of the series and parallel systems formed with components from various parametric models; see Fang and Zhang (2015), Zhao and Balakrishnan (2011), Fang and Balakrishnan (2016), Li and Li (2015), Torrado (2015), Torrado and Kochar (2015), Kundu and Chowdhury (2016), Nadarajah et al. (2017), Majumder et al. (2020) and the references therein.

Here, we investigate comparison results between the lifetimes of series and parallel systems formed with LLD samples in terms of different ordering notions such as stochastic order, hazard rate order, reversed hazard rate order, and likelihood ratio order. These orders are widely used in the literature for fair and reasonable comparison (see Shaked and Shanthikumar 2007). The rest of the paper is presented as follows. Preliminary definitions and useful lemmas can be found in Sect. 2. In Sect. 3, we discuss the comparison of lifetimes of parallel systems with heterogeneous LLD components. We also study the comparison in the case of the multiple-outlier LLD model. In Sect. 4, ordering properties are discussed for the lifetimes of series systems with heterogeneous LLD components.

Throughout this article, 'increasing' and 'decreasing' mean 'nondecreasing' and 'nonincreasing,' respectively, and the notation $f(x) \overset{\text{sign}}{=} g(x)$ implies that $f(x)$ and $g(x)$ are equal in sign.

2 Notations, Definitions, and Preliminaries

Here, we review some definitions and various notions of stochastic orders and majorization concepts.

Definition 1 (Shaked and Shanthikumar 2007) Let X and Y be two absolutely continuous r.v.'s with cumulative distribution functions (cdfs) $F(\cdot)$ and $G(\cdot)$, survival functions $\bar{F}(\cdot)$ and $\bar{G}(\cdot)$, pdfs $f(\cdot)$ and $g(\cdot)$, hazard rates $h_F(\cdot)$ and $h_G(\cdot)$, and reverse hazard rate functions $r_F(\cdot)$ and $r_G(\cdot)$, respectively.

(i) If $\bar{F}(x) \le \bar{G}(x)$ for all $x \ge 0$, then X is smaller than Y in the usual stochastic order, denoted by $X \le_{st} Y$.
(ii) If $\bar{G}(x)/\bar{F}(x)$ is increasing in $x \ge 0$, then X is smaller than Y in the hazard rate order, denoted by $X \le_{hr} Y$.
(iii) If $G(x)/F(x)$ is increasing in $x \ge 0$, then X is smaller than Y in the reversed hazard rate order, denoted by $X \le_{rh} Y$.
(iv) If $g(x)/f(x)$ is increasing in $x \ge 0$, then X is smaller than Y in the likelihood ratio order, denoted by $X \le_{lr} Y$.

From Shaked and Shanthikumar (2007), it is well established that

$$X \le_{lr} Y \implies X \le_{hr} Y \implies X \le_{st} Y$$

and

$$X \le_{lr} Y \implies X \le_{rh} Y \implies X \le_{st} Y$$

but the opposite implications do not hold in general. Also, $X \le_{hr} Y \nLeftrightarrow X \le_{rh} Y$.

The notion of majorization is a key concept in the theory of stochastic inequalities. Let $(x_{(1)}, x_{(2)}, \ldots, x_{(n)})$ denote the components of the vector $\mathbf{x} = (x_1, x_2, \ldots, x_n)$

arranged in ascending order. Let I^n be a subset of the n-dimensional Euclidean space $\mathbb{R}^n$, where $I \subseteq \mathbb{R}$ and $\mathbf{x} = (x_1, x_2, \ldots, x_n)$ be a vector in I^n.

Definition 2 The vector $\mathbf{x}$ is said to be majorized by the vector $\mathbf{y}$, denoted by $\mathbf{x} \overset{m}{\preceq} \mathbf{y}$, if

$$\sum_{i=1}^{j} x_{(i)} \geq \sum_{i=1}^{j} y_{(i)} \quad \text{for } j = 1, \ldots, n-1$$

and

$$\sum_{i=1}^{n} x_{(i)} = \sum_{i=1}^{n} y_{(i)}.$$

In addition, the vector $\mathbf{x}$ is said to be weakly supermajorized by the vector $\mathbf{y}$, denoted by $\mathbf{x} \preceq^{w} \mathbf{y}$, if

$$\sum_{i=1}^{j} x_{(i)} \geq \sum_{i=1}^{j} y_{(i)} \quad \text{for } j = 1, \ldots, n.$$

Clearly,

$$\mathbf{x} \overset{m}{\preceq} \mathbf{y} \Longrightarrow \mathbf{x} \preceq^{w} \mathbf{y}. \tag{2.1}$$

Definition 3 A real-valued function $\phi : \mathbb{R}^n \to \mathbb{R}$ is said to be Schur-convex (Schur-concave) on $\mathbb{R}^n$ if $\mathbf{x} \overset{m}{\preceq} \mathbf{y}$ implies $\phi(\mathbf{x}) \leq (\geq)\, \phi(\mathbf{y})$ for all $\mathbf{x}, \mathbf{y} \in \mathbb{R}^n$.

An extensive and comprehensive review on majorization can be found in Marshall et al. (2011).

We now introduced some well-known results which will be used in the subsequent sections to prove our main theorems.

Lemma 1 *A real-valued function ψ on I^n has the property*

$$\psi(\mathbf{x}) \leq \psi(\mathbf{y}) \quad \text{whenever} \quad \mathbf{x} \preceq^{w} \mathbf{y}$$

if and only if ψ is decreasing and Schur-convex on I^n.

Lemma 2 (Schur–Ostrowski criterion). *A continuously differentiable function $\phi : I^n \to \mathbb{R}$ is Schur-convex (Schur-concave) if and only if ϕ is symmetric and*

$$(x_i - x_j)\left(\frac{\partial \phi(\mathbf{x})}{\partial x_i} - \frac{\partial \phi(\mathbf{x})}{\partial x_j}\right) \geq (\leq) 0$$

for all $i \neq j$ and $\mathbf{x} \in I^n$.

The following two lemmas are easy to establish.

Lemma 3 *For $x \geq 0$, the function $\kappa(x) := (1 + x^{\alpha})^{-1}$ is decreasing in x for any $\alpha > 0$ and convex in x for $0 < \alpha \leq 1$. Also, the function $\tau(x) := 1 - \kappa(x)$ is concave in x for $0 < \alpha \leq 1$.*

Lemma 4 *For $x \geq 0$, the function $\varphi(x) := -\alpha x^{\alpha-1}(1 + x^{\alpha})^{-3}$ is increasing in x for $0 < \alpha \leq 1$.*

3 Order Relations for Parallel Systems

This section considers stochastic comparisons between the lifetimes of parallel systems whose components arise from two sets of heterogeneous LLD samples with a common shape parameter but different scale parameters and vice versa.

Let X_{γ_i} for $i = 1, \ldots, n$ be n independent nonnegative r.v.'s following $LLD(\alpha, \gamma_i)$ with density function given by (1.1). Let the lifetime of the parallel system formed from $X_{\gamma_1}, X_{\gamma_2}, \ldots, X_{\gamma_n}$ be $X^{\gamma}_{n:n}$. Then, its distribution and density functions are given by

$$F^{\gamma}_{n:n}(x) = \prod_{i=1}^{n} F_{\gamma_i}(x), \quad f^{\gamma}_{n:n}(x) = \prod_{i=1}^{n} F_{\gamma_i}(x) \sum_{i=1}^{n} r_{F_{\gamma_i}}(x),$$

and the corresponding reversed hazard rate function is

$$r^{\gamma}_{n:n}(x) = \frac{f^{\gamma}_{n:n}(x)}{F^{\gamma}_{n:n}(x)} = \sum_{i=1}^{n} r_{F_{\gamma_i}}(x).$$

At first, we compare two different parallel systems with common shape parameter under reversed hazard rate ordering.

Theorem 1 *For $i = 1, 2, \ldots, n$, let X_{γ_i} and X_{β_i} be two sets of independent r.v.'s such that $X_{\gamma_i} \sim LLD(\alpha, \gamma_i)$ and $X_{\beta_i} \sim LLD(\alpha, \beta_i)$ where $\gamma_i, \beta_i > 0$. Then for $0 < \alpha \leq 1$,*

$$(\gamma_1, \ldots, \gamma_n) \preceq^{w} (\beta_1, \ldots, \beta_n) \Longrightarrow X^{\gamma}_{n:n} \leq_{rh} X^{\beta}_{n:n}.$$

Proof Fix $x \geq 0$. The reversed hazard rate function of $X^{\gamma}_{n:n}$ is

$$r^{\gamma}_{n:n}(x) = \sum_{i=1}^{n} \alpha x^{-1} \left(1 + (\gamma_i x)^{\alpha}\right)^{-1} = \alpha x^{-1} \sum_{i=1}^{n} \kappa(\gamma_i x)$$

where $\kappa(x)$ is defined as in Lemma 3. From Lemma 1, it is sufficient to prove that, for every $x \geq 0$, $r^{\gamma}_{n:n}(x)$ is decreasing in each γ_i and a Schur-convex function of $(\gamma_1, \ldots, \gamma_n)$. Now from the Proposition C.1 of Marshall et al. (2011), to demonstrate the Schur-convexity of $r^{\gamma}_{n:n}(x)$, it is sufficient to prove the convexity of $\kappa(x)$. Thus, using Lemma 3 the proof follows from Definition 1.

One can have the following corollary which is an easy consequence of the relation (2.1).

Corollary 1 *For $i = 1, 2, \ldots, n$, let X_{γ_i} and X_{β_i} be two sets of independent r.v.'s such that $X_{\gamma_i} \sim LLD(\alpha, \gamma_i)$ and $X_{\beta_i} \sim LLD(\alpha, \beta_i)$ where $\gamma_i, \beta_i > 0$. Then for $0 < \alpha \leq 1$,*

$$(\gamma_1, \ldots, \gamma_n) \overset{m}{\preceq} (\beta_1, \ldots, \beta_n) \Longrightarrow X^{\gamma}_{n:n} \leq_{rh} X^{\beta}_{n:n}.$$

The above theorem ensures that for two parallel systems having independent LLD components with common shape parameter, the majorized scale parameter vector leads to corresponding system lifetime smaller in the sense of the reversed hazard rate ordering. In the following theorem, we investigate whether the systems are ordered under likelihood ratio ordering for the case $n = 2$.

Theorem 2 *For $i = 1, 2$, let X_{γ_i} and X_{β_i} be two sets of independent r.v.'s such that $X_{\gamma_i} \sim LLD(\alpha, \gamma_i)$ and $X_{\beta_i} \sim LLD(\alpha, \beta_i)$ where $\gamma_i, \beta_i > 0$. Then for $0 < \alpha \leq 1$,*

$$(\gamma_1, \gamma_2) \overset{m}{\preceq} (\beta_1, \beta_2) \Longrightarrow X^{\gamma}_{2:2} \leq_{lr} X^{\beta}_{2:2}.$$

Proof In view of Definition 1, it is enough to show that

$$\frac{f^{\beta}_{2:2}(x)}{f^{\gamma}_{2:2}(x)} = \frac{F^{\beta}_{2:2}(x)}{F^{\gamma}_{2:2}(x)} \cdot \frac{r^{\beta}_{2:2}(x)}{r^{\gamma}_{2:2}(x)} \text{ is increasing in } x. \tag{3.1}$$

From Corollary 1, we already have $F^{\beta}_{2:2}(x)/F^{\gamma}_{2:2}(x)$ is increasing in x for $0 < \alpha \leq 1$. So, (3.1) implies that it only remains to show that $\psi(x) = r^{\beta}_{2:2}(x)/r^{\gamma}_{2:2}(x)$ is increasing in x. Now the reversed hazard rate function of $X^{\beta}_{2:2}$ is given by

$$r^{\beta}_{2:2}(x) = \alpha x^{-1}\left[(1 + (\beta_1 x)^{\alpha})^{-1} + (1 + (\beta_2 x)^{\alpha})^{-1}\right].$$

Then, $\psi(x) = \dfrac{\kappa(\beta_1 x) + \kappa(\beta_2 x)}{\kappa(\gamma_1 x) + \kappa(\gamma_2 x)}$, where $\kappa(x)$ is defined as in Lemma 3. Observe that

$$\kappa'(x) = -\alpha x^{\alpha-1}(1 + x^{\alpha})^{-2} = \alpha x^{-1}\kappa(x)\eta(x)$$

where $\eta(x) = \kappa(x) - 1$. Differentiating $\psi(x)$ with respect to x, we get

$$\begin{aligned}
\psi'(x) &\overset{\text{sign}}{=} \left[\kappa'(\beta_1 x) + \kappa'(\beta_2 x)\right]\left[\kappa(\gamma_1 x) + \kappa(\gamma_2 x)\right] - \left[\kappa(\beta_1 x) + \kappa(\beta_2 x)\right]\left[\kappa'(\gamma_1 x) + \kappa'(\gamma_2 x)\right] \\
&\overset{\text{sign}}{=} \left[\kappa(\beta_1 x)\eta(\beta_1 x) + \kappa(\beta_2 x)\eta(\beta_2 x)\right]\left[\kappa(\gamma_1 x) + \kappa(\gamma_2 x)\right] \\
&- \left[\kappa(\beta_1 x) + \kappa(\beta_2 x)\right]\left[\kappa(\gamma_1 x)\eta(\gamma_1 x) + \kappa(\gamma_2 x)\eta(\gamma_2 x)\right]
\end{aligned}$$

Thus showing that $\psi(x)$ is increasing in x, i.e., $\psi^{'}(x) \geq 0\ \forall\ x \geq 0$, is equivalent to proving

$$\phi(\beta_1, \beta_2) = \frac{\kappa(\beta_1 x)\eta(\beta_1 x) + \kappa(\beta_2 x)\eta(\beta_2 x)}{\kappa(\beta_1 x) + \kappa(\beta_2 x)}$$

is Schur-convex in (β_1, β_2). Now, the function $\varphi(x)$ defined in Lemma 4 turns out to be $\kappa(x)\eta'(x)$, where $\kappa(x)$ and $\eta'(x)$ are defined as before. We thus have

$$\begin{aligned}\frac{\partial\phi}{\partial\beta_1} &\overset{\text{sign}}{=} \left[\kappa'(\beta_1 x)\eta(\beta_1 x) + \kappa(\beta_1 x)\eta'(\beta_1 x)\right]\left[\kappa(\beta_1 x) + \kappa(\beta_2 x)\right] \\ &\quad - \left[\kappa(\beta_1 x)\eta(\beta_1 x) + \kappa(\beta_2 x)\eta(\beta_2 x)\right]\kappa'(\beta_1 x) \\ &= \kappa'(\beta_1 x)\kappa(\beta_2 x)\left[\eta(\beta_1 x) - \eta(\beta_2 x)\right] + \varphi(\beta_1 x)\left[\kappa(\beta_1 x) + \kappa(\beta_2 x)\right].\end{aligned}$$

and

$$\frac{\partial\phi}{\partial\beta_2} \overset{\text{sign}}{=} \kappa(\beta_1 x)\kappa'(\beta_2 x)\left[\eta(\beta_2 x) - \eta(\beta_1 x)\right] + \varphi(\beta_2 x)\left[\kappa(\beta_1 x) + \kappa(\beta_2 x)\right].$$

Thus,

$$\begin{aligned}\frac{\partial\phi}{\partial\beta_1} - \frac{\partial\phi}{\partial\beta_2} &\overset{\text{sign}}{=} \left[\eta(\beta_1 x) - \eta(\beta_2 x)\right]\left[\kappa'(\beta_1 x)\kappa(\beta_2 x) + \kappa'(\beta_2 x)\kappa(\beta_1 x)\right] \\ &\quad + \left[\kappa(\beta_1 x) + \kappa(\beta_2 x)\right]\left[\varphi(\beta_1 x) - \varphi(\beta_2 x)\right].\end{aligned}$$

From Lemma 4, $\varphi(x)$ is increasing in x for $0 < \alpha \leq 1$. This together with the observation $\beta_1 \leq \beta_2$ and the facts that $\kappa(x)$ and $\eta(x)$ are decreasing functions of x yields

$$(\beta_1 - \beta_2)\left(\frac{\partial\phi}{\partial\beta_1} - \frac{\partial\phi}{\partial\beta_2}\right) \geq 0.$$

Hence, from Lemma 2 the theorem follows.

It is worth mentioning here that for $\alpha > 1$ the above result may not hold, as the next example shows.

Example 1 Let $(X_{\gamma_1}, X_{\gamma_2})$ and $(X_{\beta_1}, X_{\beta_2})$ be two sets of vectors of heterogeneous LLD r.v.'s with shape parameter $\alpha = 1.5$ and scale parameters $(\gamma_1, \gamma_2) = (0.5, 1.5)$ and $(\beta_1, \beta_2) = (0.3, 1.7)$. Then obviously $(\gamma_1, \gamma_2) \overset{m}{\preceq} (\beta_1, \beta_2)$ but $f^{\beta}_{2:2}(x)/f^{\gamma}_{2:2}(x)$ is not monotonic as is evident from Fig. 1. Hence in Theorem 2, the restriction over α is necessary to get the $\leq_{lr}$ order comparison.

Next theorem shows that the likelihood ratio order holds among two parallel systems formed with heterogeneous LLD components where heterogeneity occurs in terms of scale parameters.

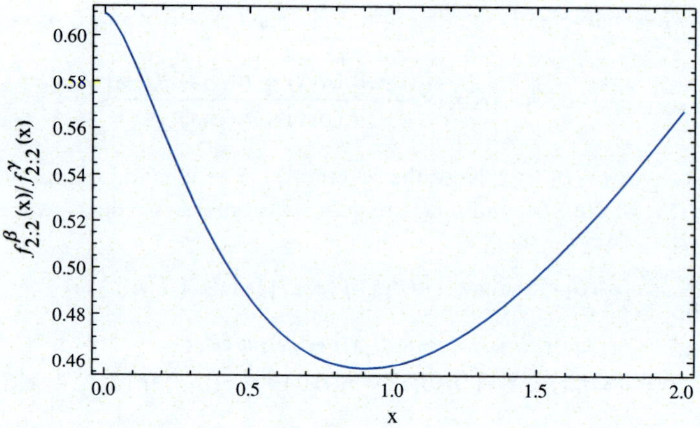

Fig. 1 Plot of $f_{2:2}^{\beta}(x)/f_{2:2}^{\gamma}(x)$ when $\alpha = 1.0$, $\gamma = (0.5, 1.5)$, and $\beta = (0.3, 1.7)$

Theorem 3 *Let X_{γ_1}, X_γ be independent r.v.'s with $X_{\gamma_1} \sim LLD(\alpha, \gamma_1)$ and $X_\gamma \sim LLD(\alpha, \gamma)$ where $\gamma_1, \gamma > 0$. Let Y_{γ^*}, Y_γ be independent r.v.'s with $Y_{\gamma^*} \sim LLD(\alpha, \gamma^*)$ and $Y_\gamma \sim LLD(\alpha, \gamma)$ where $\gamma^*, \gamma > 0$. Suppose that $\gamma^* = min(\gamma, \gamma_1, \gamma^*)$ then for any $\alpha > 0$,*

$$(\gamma_1, \gamma) \preceq^w (\gamma^*, \gamma) \Longrightarrow X_{2:2} \leq_{lr} Y_{2:2}^*.$$

Proof The reversed hazard rate function of $X_{2:2}(x)$ has the form

$$r_{2:2}(x) = \alpha x^{-1}\left[(1 + (\gamma_1 x)^\alpha)^{-1} + (1 + (\gamma x)^\alpha)^{-1}\right].$$

Let $\psi(x) = \dfrac{r_{2:2}^*(x)}{r_{2:2}(x)} = \dfrac{(1 + (\gamma x)^\alpha)^{-1} + (1 + (\gamma^* x)^\alpha)^{-1}}{(1 + (\gamma x)^\alpha)^{-1} + (1 + (\gamma_1 x)^\alpha)^{-1}}$. Now utilizing Eq. (3.1) and Theorem 1, it only remains to show that $\psi(x)$ is increasing in x, i.e., $\psi'(x) \geq 0, \ \forall x \geq 0$. Now differentiating $\psi(x)$ with respect to x and using the functions $\kappa(x)$ and $\eta(x)$ defined earlier, we get

$$\begin{aligned}
\psi'(x) \overset{\text{sign}}{=} & \left[(1 + (\gamma_1 x)^\alpha)^{-1} + (1 + (\gamma x)^\alpha)^{-1}\right]\left[-(\gamma x)^\alpha(1 + (\gamma x)^\alpha)^{-2} - (\gamma^* x)^\alpha(1 + (\gamma^* x)^\alpha)^{-2}\right] \\
- & \left[(1 + (\gamma x)^\alpha)^{-1} + (1 + (\gamma^* x)^\alpha)^{-1}\right]\left[-(\gamma x)^\alpha(1 + (\gamma x)^\alpha)^{-2} - (\gamma_1 x)^\alpha(1 + (\gamma_1 x)^\alpha)^{-2}\right] \\
= & \left[\kappa(\gamma x)\eta(\gamma x) + \kappa(\gamma^* x)\eta(\gamma^* x)\right]\left[\kappa(\gamma_1 x) + \kappa(\gamma x)\right] \\
- & \left[\kappa(\gamma x) + \kappa(\gamma^* x)\right]\left[\kappa(\gamma_1 x)\eta(\gamma_1 x) + \kappa(\gamma x)\eta(\gamma x)\right] \\
= & \ \kappa(\gamma_1 x)\kappa(\gamma^* x)\left[\eta(\gamma^* x) - \eta(\gamma_1 x)\right] + \kappa(\gamma_1 x)\kappa(\gamma x)\left[\eta(\gamma x) - \eta(\gamma_1 x)\right] \\
+ & \ \kappa(\gamma x)\kappa(\gamma^* x)\left[\eta(\gamma^* x) - \eta(\gamma x)\right].
\end{aligned}$$

Since $(\gamma_1, \gamma) \preceq^w (\gamma^*, \gamma)$ and $\gamma^* = \min(\gamma, \gamma_1, \gamma^*)$, two cases may arise:
Case I: $\gamma^* \leq \gamma \leq \gamma_1$. It can be easily seen that $\psi'(x) \geq 0$, using the facts $\kappa(x) \geq$

$0 \ \forall x \geq 0$ and $\eta(x)$ is decreasing in x.
Case II: $\gamma^* \leq \gamma_1 \leq \gamma$. Again utilizing the above facts, we have

$$\begin{aligned}\psi'(x) &\geq \kappa(\gamma_1 x)\kappa(\gamma x)\left[\eta(\gamma^* x) - \eta(\gamma_1 x)\right] + \kappa(\gamma_1 x)\kappa(\gamma x)\left[\eta(\gamma x) - \eta(\gamma_1 x)\right] \\ &\quad + \kappa(\gamma x)\kappa(\gamma_1 x)\left[\eta(\gamma^* x) - \eta(\gamma x)\right] \\ &= 2\kappa(\gamma x)\kappa(\gamma_1 x)\left[\eta(\gamma^* x) - \eta(\gamma_1 x)\right] \geq 0.\end{aligned}$$

Thus in both the cases, one has $\psi(x)$ is increasing in x. Hence, the theorem follows.

Now we establish a comparison between parallel systems based on two sets of heterogeneous LLD r.v.'s with common scale parameter and majorized shape parameters according to stochastic ordering.

Theorem 4 *For $i = 1, 2, \ldots, n$, let X_{α_i} and X_{β_i} be two sets of independent r.v.'s with $X_{\alpha_i} \sim LLD(\alpha_i, \gamma)$ and $X_{\beta_i} \sim LLD(\beta_i, \gamma)$ where $\alpha_i, \beta_i > 0$. Then for any $\gamma > 0$,*

$$(\alpha_1, \ldots, \alpha_n) \overset{m}{\preceq} (\beta_1, \ldots, \beta_n) \Longrightarrow X^{\alpha}_{n:n} \leq_{st} X^{\beta}_{n:n}.$$

Proof The distribution function of $X^{\alpha}_{n:n}$ is

$$F^{\alpha}_{n:n}(x) = \prod_{i=1}^{n} F_{\alpha_i}(x) = \prod_{i=1}^{n} (\gamma x)^{\alpha_i}(1 + (\gamma x)^{\alpha_i})^{-1} = \prod_{i=1}^{n} \zeta_{\gamma x}(\alpha_i)$$

where $\zeta_x(\alpha) = x^{\alpha}/(1 + x^{\alpha})$, $x, \alpha > 0$. From Definition 1, we have to show that $F^{\alpha}_{n:n}(x)$ is Schur-concave in $(\alpha_1, \ldots, \alpha_n)$. Proposition E.1. of Marshall et al. (2011) implies that it is sufficient to check the concavity of $\log_e \zeta_x(\alpha)$, in order to establish the Schur-concavity of $F^{\alpha}_{n:n}(x)$. Observe that the function $\log_e \zeta_x(\alpha)$ is concave in α for all $\gamma > 0$. Hence, $F^{\alpha}_{n:n}(x)$ is Schur-concave in $(\alpha_1, \ldots, \alpha_n)$.

Next, we investigate whether the above result can be generalized to the case of reversed hazard rate ordering. Consider the following example:

Example 2 Let $X_{\alpha_i} \sim LLD(\alpha_i, \gamma)$ and $X_{\beta_i} \sim LLD(\beta_i, \gamma)$ for $i = 1, 2$, where $(\alpha_1, \alpha_2) = (2.5, 1.5)$ and $(\beta_1, \beta_2) = (1, 3)$ with common scale parameter $\gamma = 2$. Obviously $(\alpha_1, \alpha_2) \overset{m}{\preceq} (\beta_1, \beta_2)$ but $X^{\alpha}_{2:2} \nleq_{rh} X^{\beta}_{2:2}$ which can be easily verified by the plot of corresponding reversed hazard rate functions in Fig. 2.

Next, we investigate the likelihood ratio ordering on maximum-order statistics arising from multiple-outlier LLD samples. Here, it is pertinent to mention that a multiple-outlier model is a set of independent r.v.'s $X_1, X_2, \ldots, X_n$ such that $X_i \overset{st}{=} X$, $i = 1, 2, \ldots, p$ and $X_i \overset{st}{=} Y$, $i = p + 1, p + 2, \ldots, p + q = n$ where $1 \leq p < n$ and $X_i \overset{st}{=} X$ means that X_i and X are identically distributed. In summary, the set of r.v.'s $X_1, X_2, \ldots, X_n$ is said to constitute a multiple-outlier model if two sets of r.v.'s

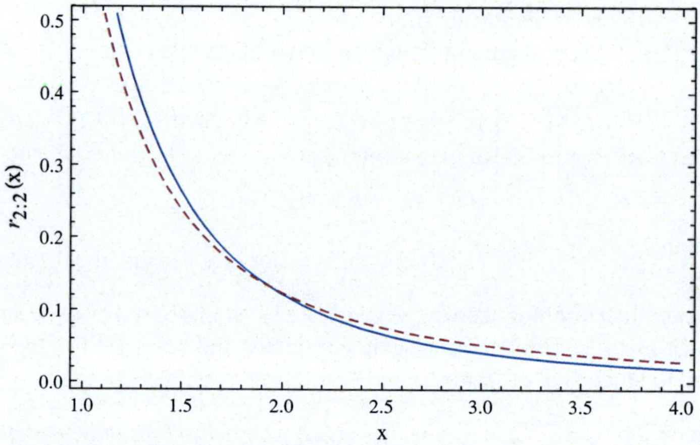

Fig. 2 Plot of the reversed hazard rate function of $X^{\alpha}_{2:2}$ (continuous line) and $X^{\beta}_{2:2}$ (dashed line) when $\gamma = 2$, $(\alpha_1, \alpha_2) = (2.5, 1.5)$ and $(\beta_1, \beta_2) = (1, 3)$

$X_1, X_2, \ldots, X_p$ and $X_{p+1}, X_{p+2}, \ldots, X_{p+q}$ are homogeneous among themselves and heterogeneous between themselves. For more details on multiple-outlier models, see Balakrishnan (2007).

The following two theorems present versions of Theorems 2 and 3 in the context of multiple-outlier models.

Theorem 5 *Let $X_1, X_2, \ldots, X_n$ be independent r.v.'s following the multiple-outlier LLD model such that $X_i \sim LLD(\alpha, \gamma_1)$ for $i = 1, 2, \ldots, p$ and $X_j \sim LLD(\alpha, \gamma_2)$ for $j = p+1,\ p+2, \ldots, n$ with $\gamma_1, \gamma_2 > 0$. Let $Y_1, Y_2, \ldots, Y_n$ be another set of independent r.v.'s following the multiple-outlier LLD model such that $Y_i \sim LLD(\alpha, \beta_1)$ for $i = 1, 2, \ldots, p$ and $Y_j \sim LLD(\alpha, \beta_2)$ for $j = p+1, p+2, \ldots, n$ with $\beta_1, \beta_2 > 0$. Then for $0 < \alpha \leq 1$,*

$$(\underbrace{\gamma_1, \ldots, \gamma_1}_{p}, \underbrace{\gamma_2, \ldots, \gamma_2}_{q}) \overset{m}{\preceq} (\underbrace{\beta_1, \ldots, \beta_1}_{p}, \underbrace{\beta_2, \cdots, \beta_2}_{q}) \Longrightarrow X^{\gamma}_{n:n} \leq_{lr} Y^{\beta}_{n:n} \text{ where } p + q = n.$$

Proof In view of Theorem 2, an equivalent form of (3.1) for this model enables us to complete the proof by simply showing that $r^{\beta}_{n:n}(x)/r^{\gamma}_{n:n}(x)$ is increasing in x. Here, the reversed hazard rate of $Y^{\beta}_{n:n}$ is

$$r^{\beta}_{n:n}(x) = \alpha x^{-1}\left[p(1 + (\beta_1 x)^{\alpha})^{-1} + q(1 + (\beta_2 x)^{\alpha})^{-1}\right]$$

where $p + q = n$. Then,

$$\psi(x) = \frac{r^{\beta}_{n:n}(x)}{r^{\gamma}_{n:n}(x)} = \frac{p\kappa(\beta_1 x) + q\kappa(\beta_2 x)}{p\kappa(\gamma_1 x) + q\kappa(\gamma_2 x)}$$

where $\kappa(x)$ is defined as in Lemma 3. Note that, for $x \geq 0$

$$\begin{aligned}\psi'(x) &\overset{\text{sign}}{=} \left[p\kappa(\gamma_1 x) + q\kappa(\gamma_2 x)\right]\left[p\kappa'(\beta_1 x) + q\kappa'(\beta_2 x)\right] \\ &- \left[p\kappa(\beta_1 x) + q\kappa(\beta_2 x)\right]\left[p\kappa'(\gamma_1 x) + q\kappa'(\gamma_2 x)\right] \\ &\overset{\text{sign}}{=} \left[p\kappa(\beta_1 x)\eta(\beta_1 x) + q\kappa(\beta_2 x)\eta(\beta_2 x)\right]\left[p\kappa(\gamma_1 x) + q\kappa(\gamma_2 x)\right] \\ &- \left[p\kappa(\beta_1 x) + q\kappa(\beta_2 x)\right]\left[p\kappa(\gamma_1 x)\eta(\gamma_1 x) + q\kappa(\gamma_2 x)\eta(\gamma_2 x)\right]\end{aligned}$$

where $\eta(x)$ is defined as in the proof of Theorem 2. To show $\psi(x)$ is increasing in x, i.e.,

$$\frac{p\kappa(\beta_1 x)\eta(\beta_1 x) + q\kappa(\beta_2 x)\eta(\beta_2 x)}{p\kappa(\beta_1 x) + q\kappa(\beta_2 x)} \geq \frac{p\kappa(\gamma_1 x)\eta(\gamma_1 x) + q\kappa(\gamma_2 x)\eta(\gamma_2 x)}{p\kappa(\gamma_1 x) + q\kappa(\gamma_2 x)},$$

it is sufficient to show that the function $\phi(\beta_1, \beta_2) = \dfrac{p\kappa(\beta_1 x)\eta(\beta_1 x) + q\kappa(\beta_2 x)\eta(\beta_2 x)}{p\kappa(\beta_1 x) + q\kappa(\beta_2 x)}$ is Schur-convex in (β_1, β_2).
Now, differentiating $\phi(\beta_1, \beta_2)$ with respect to β_1, we obtain

$$\begin{aligned}\frac{\partial\phi}{\partial\beta_1} &\overset{\text{sign}}{=} \left[\kappa'(\beta_1 x)\eta(\beta_1 x) + \kappa(\beta_1 x)\eta'(\beta_1 x)\right]\left[p\kappa(\beta_1 x) + q\kappa(\beta_2 x)\right] \\ &- \left[p\kappa(\beta_1 x)\eta(\beta_1 x) + q\kappa(\beta_2 x)\eta(\beta_2 x)\right]\kappa'(\beta_1 x) \\ &= q\kappa'(\beta_1 x)\kappa(\beta_2 x)\left[\eta(\beta_1 x) - \eta(\beta_2 x)\right] + \varphi(\beta_1 x)\left[p\kappa(\beta_1 x) + q\kappa(\beta_2 x)\right]\end{aligned}$$

where $\varphi(x)$ is as defined in Theorem 2. By interchanging β_1 and β_2, we obtain

$$\frac{\partial\phi}{\partial\beta_2} \overset{\text{sign}}{=} p\kappa(\beta_1 x)\kappa'(\beta_2 x)\left[\eta(\beta_2 x) - \eta(\beta_1 x)\right] + \varphi(\beta_2 x)\left[p\kappa(\beta_1 x) + q\kappa(\beta_2 x)\right].$$

Now,

$$\begin{aligned}\frac{\partial\phi}{\partial\beta_1} - \frac{\partial\phi}{\partial\beta_2} &\overset{\text{sign}}{=} \left[\eta(\beta_1 x) - \eta(\beta_2 x)\right]\left[q\kappa'(\beta_1 x)\kappa(\beta_2 x) + p\kappa'(\beta_2 x)\kappa(\beta_1 x)\right] \\ &+ \left[p\kappa(\beta_1 x) + q\kappa(\beta_2 x)\right]\left[\varphi(\beta_1 x) - \varphi(\beta_2 x)\right]\end{aligned}$$

Since $\beta_1 \leq \beta_2$ and $\kappa(x)$ and $\eta(x)$ are decreasing in x and $\varphi(x)$ is increasing in x, we have, for $0 < \alpha \leq 1$

$$(\beta_1 - \beta_2)\left(\frac{\partial\phi}{\partial\beta_1} - \frac{\partial\phi}{\partial\beta_2}\right) \geq 0.$$

Hence, $\phi(\beta_1, \beta_2)$ is Schur-convex in (β_1, β_2) and consequently the theorem follows.

Theorem 6 *Let $X_1, X_2, \ldots, X_n$ be independent r.v.'s following the multiple-outlier LLD model such that $X_i \sim LLD(\alpha, \gamma_1)$ for $i = 1, 2, \ldots, p$ and $X_j \sim LLD(\alpha, \gamma)$ for $j = p+1, p+2, \ldots, n$ with $\gamma_1, \gamma > 0$. Let $Y_1, Y_2, \cdots, Y_n$ be another set of inde-*

pendent r.v.'s following the multiple-outlier LLD model such that $Y_i \sim LLD(\alpha, \gamma^*)$ for $i = 1, 2, \ldots, p$ and $Y_j \sim LLD(\alpha, \gamma)$ for $j = p+1, p+2, \ldots, n$ with $\gamma^*, \gamma > 0$. Suppose that $\gamma^* = min(\gamma, \gamma_1, \gamma^*)$ then for any $\alpha > 0$,

$$(\underbrace{\gamma_1, \ldots, \gamma_1}_{p}, \underbrace{\gamma, \ldots, \gamma}_{q}) \overset{w}{\preceq} (\underbrace{\gamma^*, \ldots, \gamma^*}_{p}, \underbrace{\gamma, \ldots, \gamma}_{q}) \Longrightarrow \frac{r^*_{n:n}(x)}{r_{n:n}(x)} \text{ is increasing in } x,$$

where $p + q = n$.

Proof The reversed hazard function of $X_{n:n}$ is

$$r_{n:n}(x) = \frac{\alpha}{x}\left[\frac{p}{1+(\gamma_1 x)^\alpha} + \frac{q}{1+(\gamma x)^\alpha}\right] \text{ where } p+q=n.$$

Let $\psi(x) = \dfrac{r^*_{n:n}(x)}{r_{n:n}(x)} = \dfrac{q(1+(\gamma x)^\alpha)^{-1} + p(1+(\gamma^* x)^\alpha)^{-1}}{q(1+(\gamma x)^\alpha)^{-1} + p(1+(\gamma_1 x)^\alpha)^{-1}}$. To show $\psi(x)$ is increasing in x, we consider

$$\begin{aligned}
\psi'(x) \overset{\text{sign}}{=} & \left[\frac{p}{1+(\gamma_1 x)^\alpha} + \frac{q}{1+(\gamma x)^\alpha}\right]\left[\frac{-q(\gamma x)^\alpha}{(1+(\gamma x)^\alpha)^2} + \frac{-p(\gamma^* x)^\alpha}{(1+(\gamma^* x)^\alpha)^2}\right] \\
& - \left[\frac{q}{1+(\gamma x)^\alpha} + \frac{p}{1+(\gamma^* x)^\alpha}\right]\left[\frac{-q(\gamma x)^\alpha}{(1+(\gamma x)^\alpha)^2} + \frac{-p(\gamma_1 x)^\alpha}{(1+(\gamma_1 x)^\alpha)^2}\right] \\
& = \left[q\kappa(\gamma x)\eta(\gamma x) + p\kappa(\gamma^* x)\eta(\gamma^* x)\right]\left[p\kappa(\gamma_1 x) + q\kappa(\gamma x)\right] \\
& - \left[q\kappa(\gamma x) + p\kappa(\gamma^* x)\right]\left[p\kappa(\gamma_1 x)\eta(\gamma_1 x) + q\kappa(\gamma x)\eta(\gamma x)\right] \\
& = p^2\kappa(\gamma_1 x)\kappa(\gamma^* x)\left[\eta(\gamma^* x) - \eta(\gamma_1 x)\right] + pq\kappa(\gamma_1 x)\kappa(\gamma x)\left[\eta(\gamma x) - \eta(\gamma_1 x)\right] \\
& + pq\kappa(\gamma x)\kappa(\gamma^* x)\left[\eta(\gamma^* x) - \eta(\gamma x)\right].
\end{aligned}$$

Now using the facts that $\eta(x)$ is decreasing in x, $\kappa(x) \geq 0 \; \forall x > 0$ and $\gamma^* = min(\gamma, \gamma_1, \gamma^*)$, it is easy to show the following: If $\gamma^* \leq \gamma \leq \gamma_1$, then $\psi'(x) \geq 0 \; \forall x > 0$. Also, if $\gamma^* \leq \gamma_1 \leq \gamma$, we have

$$\begin{aligned}
\psi'(x) \geq & \; p^2\kappa(\gamma_1 x)\kappa(\gamma x)\left[\eta(\gamma^* x) - \eta(\gamma_1 x)\right] + pq\kappa(\gamma_1 x)\kappa(\gamma x)\left[\eta(\gamma x) - \eta(\gamma_1 x)\right] \\
& + pq\kappa(\gamma_1 x)\kappa(\gamma x)\left[\eta(\gamma^* x) - \eta(\gamma x)\right] \\
& = np\kappa(\gamma x)\kappa(\gamma_1 x)\left[\eta(\gamma^* x) - \eta(\gamma_1 x)\right] \geq 0.
\end{aligned}$$

Thus in both the cases, $\psi'(x) \geq 0 \; \forall x > 0$ and the theorem follows.

Observe that if $(\gamma_1, \gamma) \preceq^w (\gamma^*, \gamma)$ where $\gamma^* = \min(\gamma, \gamma_1, \gamma^*)$ then the parallel system formed by $LLD(\alpha, \gamma_1)$ and $LLD(\alpha, \gamma)$ has the smaller lifetime than the system formed with $LLD(\alpha, \gamma^*)$ and $LLD(\alpha, \gamma)$ in the reverse hazard rate sense for any shape parameter $\alpha > 0$. Using this fact together with the result in Theorem 1.C.4. of Shaked and Shanthikumar (2007), one can get the following result.

Theorem 7 *Let $X_1, X_2, \ldots, X_n$ be independent r.v.'s following the multiple-outlier LLD model such that $X_i \sim LLD(\alpha, \gamma_1)$ for $i = 1, 2, \ldots, p$ and $X_j \sim LLD(\alpha, \gamma)$ for $j = p+1, p+2, \ldots, n$ with $\gamma_1, \gamma > 0$. Let $Y_1, Y_2, \ldots, Y_n$ be another set of independent r.v.'s following the multiple-outlier LLD model such that $Y_i \sim LLD(\alpha, \gamma^*)$ for $i = 1, 2, \ldots, p$ and $Y_j \sim LLD(\alpha, \gamma)$ for $j = p+1, p+2, \ldots, n$ with $\gamma^*, \gamma > 0$. Suppose that $\gamma^* = min(\gamma, \gamma_1, \gamma^*)$ then for any $\alpha > 0$,*

$$(\underbrace{\gamma_1, \ldots, \gamma_1}_{p}, \underbrace{\gamma, \ldots, \gamma}_{q}) \overset{w}{\preceq} (\underbrace{\gamma^*, \ldots, \gamma^*}_{p}, \underbrace{\gamma, \ldots, \gamma}_{q}) \Longrightarrow X_{n:n} \leq_{lr} Y^*_{n:n} \text{ where } p+q=n.$$

4 Order Relations for Series System

In this section, our main aim is to compare two series systems formed with independent heterogeneous LLD samples either having common shape parameter but different scale parameters or conversely.

Let $X^{\gamma}_{1:n}$ denote the lifetime of the series system formed with n independent non-negative r.v.'s $X_{\gamma_1}, X_{\gamma_2}, \ldots, X_{\gamma_n}$, where each $X_{\gamma_i} \sim LLD(\alpha, \gamma_i)$. Then, its survival and density functions are given by

$$\bar{F}^{\gamma}_{1:n}(x) = \prod_{i=1}^{n} \bar{F}_{\gamma_i}(x), \quad f^{\gamma}_{1:n}(x) = \prod_{i=1}^{n} \bar{F}_{\gamma_i}(x) \sum_{i=1}^{n} h_{F_{\gamma_i}}(x),$$

and the corresponding hazard rate function is

$$h^{\gamma}_{1:n}(x) = \frac{f^{\gamma}_{1:n}(x)}{\bar{F}^{\gamma}_{1:n}(x)} = \sum_{i=1}^{n} h_{F_{\gamma_i}}(x).$$

The following theorem shows that under a certain condition on the shape parameter, one can compare the lifetimes of two series systems with independent LLD components according to hazard rate ordering.

Theorem 8 *For $i = 1, 2, \ldots, n$, let X_{λ_i} and X_{β_i} be two sets of independent r.v.'s with $X_{\gamma_i} \sim LLD(\alpha, \gamma_i)$ and $X_{\beta_i} \sim LLD(\alpha, \beta_i)$ where $\gamma_i, \beta_i > 0$. Then for $0 < \alpha \leq 1$,*

$$(\gamma_1, \ldots, \gamma_n) \overset{m}{\preceq} (\beta_1, \ldots, \beta_n) \Longrightarrow X^{\gamma}_{1:n} \leq_{hr} X^{\beta}_{1:n}.$$

Proof Fix $x \geq 0$. The hazard rate function of $X^{\gamma}_{1:n}$ is

$$h^{\gamma}_{1:n}(x) = \sum_{i=1}^{n} \alpha x^{-1} (\gamma_i x)^{\alpha} (1 + (\gamma_i x)^{\alpha})^{-1} = \sum_{i=1}^{n} \alpha x^{-1} \tau(\gamma_i x)$$

where $\tau(x)$ is as defined in Lemma 3 and is concave in x for $0 < \alpha \leq 1$. It follows from Proposition C.1. of Marshall et al. (2011) that $\sum_{i=1}^{n} \tau(\gamma_i x)$ is Schur-concave. This completes the proof.

From the theory of stochastic ordering, we have $\leq_{rh} \Longrightarrow \leq_{st}$. Thus from the above theorem, it is clear that the result is also valid in the sense of stochastic ordering. The next question that arises naturally is whether the comparison can be extended to likelihood ratio ordering, i.e., if a version of Theorem 3 for comparison in the sense of likelihood ratio ordering is valid in the context of series systems. The following example gives the answer in the negative.

Example 3 Let $X_{\gamma_i} \sim LLD(\alpha, \gamma_i)$ and $X_{\beta_i} \sim LLD(\alpha, \beta_i)$ for $i = 1, 2$ where the scale parameters are $(\gamma_1, \gamma_2) = (0.5, 1.5)$ and $(\beta_1, \beta_2) = (0.3, 1.7)$, respectively. Now the plot of $f_{1:n}^{\beta}(x)/f_{1:n}^{\gamma}(x)$ for the common shape parameters $\alpha = 0.5$ and $\alpha = 1.5$ is given in Figs. 3a, b, respectively. Obviously in both the cases, $(\gamma_1, \gamma_2) \overset{m}{\preceq} (\beta_1, \beta_2)$ holds but $X_{2:2}^{\gamma} \nleq_{lr} X_{2:2}^{\beta}$ since in both the cases $f_{1:2}^{\beta}(x)/f_{1:2}^{\gamma}(x)$ is not a monotonic function.

Now we consider series systems having heterogeneous LLD components with common scale parameter and different shape parameters (which are also majorized) and investigate similar results.

Theorem 9 *For $i = 1, 2, \ldots, n$, let X_{α_i} and X_{β_i} be two sets of independent r.v.'s with $X_{\alpha_i} \sim LLD(\alpha_i, \gamma)$ and $X_{\beta_i} \sim LLD(\beta_i, \gamma)$ where $\alpha_i, \beta_i > 0$. Then for any $\gamma > 0$,*

$$(\alpha_1, \ldots, \alpha_n) \overset{m}{\preceq} (\beta_1, \ldots, \beta_n) \Longrightarrow X_{1:n}^{\alpha} \geq_{st} X_{1:n}^{\beta}.$$

Proof The survival function of $X_{1:n}^{\alpha}$ is

$$\bar{F}_{1:n}^{\alpha}(x) = \prod_{i=1}^{n} \bar{F}_{\alpha_i}(x) = \prod_{i=1}^{n} (1 + (\gamma x)^{\alpha_i})^{-1} = \prod_{i=1}^{n} \upsilon_{\gamma x}(\alpha_i)$$

where $\upsilon_{\gamma x}(\alpha_i) = (1 + (\gamma x)^{\alpha_i})^{-1}$. To establish the result, it is enough to show that $\bar{F}_{1:n}^{\alpha}(x)$ is Schur-concave in $(\alpha_1, \ldots, \alpha_n)$. Observe that the function $\log_e \upsilon_{\lambda x}(\alpha)$ is concave in α for all $\gamma > 0$. Then, an argument similar to that of Theorem 4 yields the result.

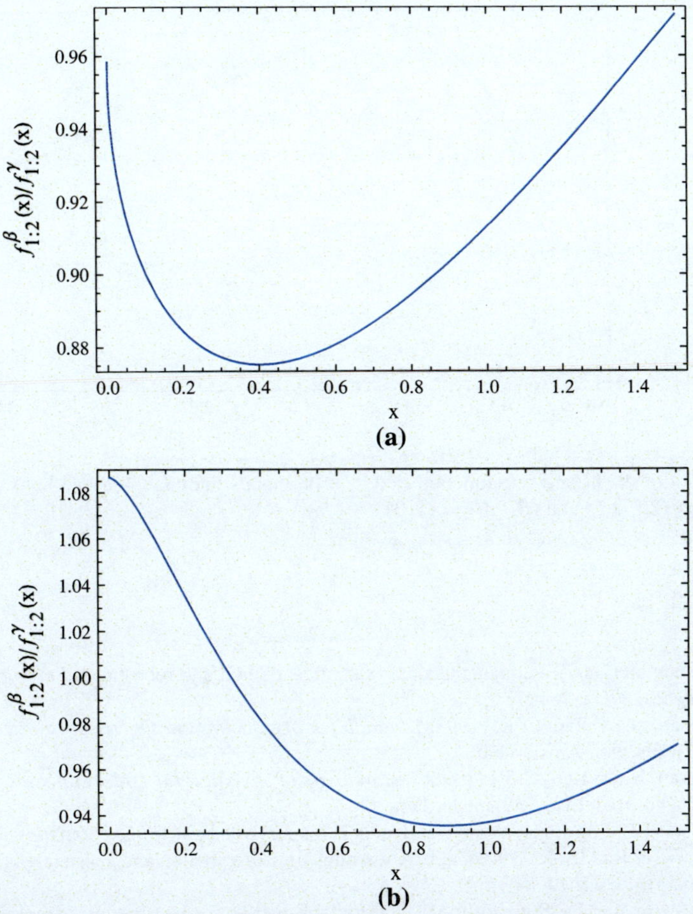

Fig. 3 Plot of $f^{\beta}_{1:2}(x)/f^{\gamma}_{1:2}(x)$ when $\alpha = 0.5$ (Sub-Fig. **a**) and $\alpha = 1.5$ (Sub-Fig. **b**) for $= (0.5, 1.5)$, and **fi** $= (0.3, 1.7)$

In this type of series system model when we compare further, the following example illustrates that no such comparison can be made in the sense of hazard rate ordering.

Example 4 Figure 4 illustrates that stochastic comparison between lifetimes of two series systems $X^{\alpha}_{1:2}$ and $X^{\beta}_{1:2}$ with LLD components having common scale parameter $\gamma = 1$ and majorized shape parameters $(\alpha_1, \alpha_2) = (2.5, 1.5)$ and $(\beta_1, \beta_2) = (1, 3)$ is not ordered in the sense of hazard rate ordering.

Acknowledgements We thank the anonymous reviewer for his/her helpful comments which have substantially improved the presentation of the paper. The authors are also grateful to Prof. Arnab K. Laha, IIMA, for his constant encouragement and words of advice.

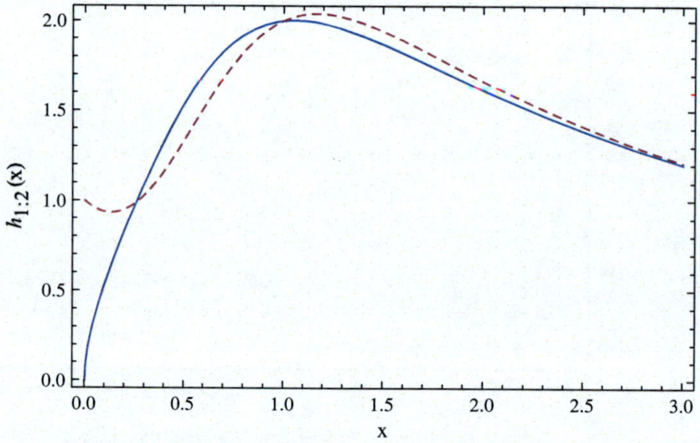

Fig. 4 Plot of the hazard rate function of $X_{1:2}^{\alpha}$ (continuous line) $X_{1:2}^{\beta}$ (dashed line) when $\gamma = 1$, $(\alpha_1, \alpha_2) = (2.5, 1.5)$ and $(\beta_1, \beta_2) = (1, 3)$

References

Balakrishnan, N. (2007). Permanents, order statistics, outlier, and robustness. *Revista matemática complutense*, *20*(1), 7–107.

Balakrishnan, N., & Rao, C. R. (1998a). *Handbook of statistics, in: Order statistics: Applications* (Vol. 17). Elsevier, Amsterdam.

Balakrishnan, N., & Rao, C. R. (1998b). *Handbook of statistics, in: Order statistics: Theory and methods* (Vol. 16). Elsevier, Amsterdam.

Bennet, S. (1983). Log-logistic regression models for survival data. *Applied Statistics*, *32*, 165–171.

Block, H. W., & Rao, B. R. (1973). A beta warning-time distribution and a distended beta distribution. *Sankhya B*, *35*, 79–84.

Fahim, A., & Smail, M. (2006). Fitting the log-logistic distribution by generalized moments. *Journal of Hydrology*, *328*, 694–703.

Fang, L., & Balakrishnan, N. (2016). Ordering results for the smallest and largest order statistics from independent heterogeneous exponential-Weibull random variables. *Statistics*, *50*(6), 1195–1205.

Fang, L., & Zhang, X. (2015). Stochastic comparisons of parallel systems with exponentiated Weibull components. *Statistics and Probability Letters*, *97*, 25–31.

Fisk, P. (1961). The graduation of income distributions. *Econometrica*, *29*, 171–185.

Gago-Benítez, A., Fernández-Madrigal, J.-A., & Cruz-Martín, A. (2013). Log-logistic modelling of sensory flow delays in networked telerobots. *IEEE Sensors*, *13*(8), 2944–2953.

Kundu, A., & Chowdhury, S. (2016). Ordering properties of order statistics from heterogeneous exponentiated Weibull models. *Statistics and Probability Letters*, *114*, 119–127.

Lawless, J. F. (1986). A note on lifetime regression models. *Biometrika*, *73*(2), 509–512.

Li, C., & Li, X. (2015). Likelihood ratio order of sample minimum from heterogeneous Weibull random variables. *Statistics and Probability Letters*, *97*, 46–53.

Majumder, P., Ghosh, S., & Mitra, M. (2020). Ordering results of extreme order statistics from heterogeneous Gompertz-Makeham random variables. *Statistics*, *54*(3), 595–617.

Malik, H. (1967). Exact distribution of the quotient of independent generalized Gamma variables. *Canadian Mathematical Bulletin*, *10*, 463–466.

Marshall, A., Olkin, I., & Arnold, B. C. (2011). *Inequalities: Theory of majorization and its applications*. New York: Springer series in Statistics.

Mielke, P. W., & Johnson, E. (1973). Three-parameter Kappa distribution maximum likelihood estimates and likelihood ratio tests. *Monthly Weather Review*, *101*, 701–709.

Nadarajah, S., Jiang, X., & Chu, J. (2017). Comparisons of smallest order statistics from Pareto distributions with different scale and shape parameters. *Annals of Operations Research*, *254*, 191–209.

Shaked, M., & Shanthikumar, J. (2007). *Stochastic orders*. New York: Springer.

Shoukri, M. M., Mian, I. U. M., & Tracy, D. S. (1988). Sampling properties of estimators of the log-logistic distribution with application to Canadian precipitation data. *The Canadian Journal of Statistics*, *16*(3), 223–236.

Torrado, N. (2015). Comparisons of smallest order statistics from Weibull distributions with different scale and shape parameters. *Journal of the Korean Statistical Society*, *44*, 68–76.

Torrado, N., & Kochar, S. C. (2015). Stochastic order relations among parallel systems from Weibull distributions. *Journal of Applied Probability*, *52*, 102–116.

Zhao, P., & Balakrishnan, N. (2011). New results on comparison of parallel systems with heterogeneous Gamma components. *Statistics and Probability Letters*, *81*, 36–44.

Stacking with Dynamic Weights on Base Models

Biswaroop Mookherjee and Abhishek Halder

1 Introduction

This paper is of methodological development of two new ways of stacking. Stacking is an ensemble technique which combines different classification techniques aiming correct classification rate higher than what the base techniques provide. Suppose, we have applied two classification techniques named CT1 and CT2 on a dataset where the objective is binary classification. Say, the base techniques CT1 and CT2 provide correct classifications of $x\%$ and $y\%$, respectively. Then, stacking is expected to provide correct classification higher than maximum of $x\%$ and $y\%$. We have proposed two new methods of stacking which perform better than the conventional way.

2 Literature Review

Stacking appears in the papers by Wolpert [1] and Breiman [2]. It is widely used by machine learning practitioners to get better classification by creating ensemble of multiple models based on different classification techniques.

In the conventional way, stacking is done by running a classification technique with outputs of the base (or primary) learners as independent variables. The target variable is kept same. Hence, one can think the overall structure as function of functions. Classification techniques are run on the dataset, and prediction of the classes is obtained. Such predicted classes go as independent variables in another model while the target variable remains same. Usually, the second-level modelling

B. Mookherjee (✉) · A. Halder
Tata Consultancy Services (TCS), Kolkata, India
e-mail: biswaroop9000@gmail.com

A. Halder
e-mail: abhishek.halder171@gmail.com

A. K. Laha (ed.), *Applied Advanced Analytics*, Springer Proceedings in Business and Economics, https://doi.org/10.1007/978-981-33-6656-5_15

is done using logistic regression. Finally, classification happens by the whole nested structure.

Stacking does not necessarily do well always because of its rigid structure of applying same set of weights on the base learners in all parts of the data. The weights are obtained from the second level of model. We have not found any procedure of stacking where the weights given on base learners vary in different parts of the data considering the performance of the base learners in different parts. Hence, we have developed methods which are narrated in Sects. 4 and 5.

3 Stacking by Conventional Way

In this conventional way, different models are prepared using different techniques. Then, the predicted classes by different models are used as predictors along with the observed class as target to run the upper-level model.

3.1 Steps of Stacking by Conventional Way

Step 1 Run k number of classification techniques $T1, T2, \ldots, Tk$ and build models.

Step 2 Take predicted classes as predictors along with the observed classes as values of the target variable

Step 3 Run a classification technique which is usually logistic regression on the dataset prepared in Step 2 to get a model

Step 4 The new observations are passed through the models prepared in Step 1 and 'k' number of outputs are obtained for each observation as 'k' number of techniques are there.

Step 5 The k outputs obtained from Step 4 are passed through the model prepared at Step 3 to get the final class prediction.

The *R* programming language functions that are used are 'glm' for logistic regression, 'lda' for linear discriminant analysis, rpart for decision tree, trainControl and train for k-nearest neighbors. Stacking by the conventional way is done by logistic regression.

4 Proposed Method I: Stacking Using Neighbourhood-Based Dynamic Weights

We propose a new way of stacking where the base learners do not get weights from the second level model. Such weights are applied on all points over the whole dataset. Our method is dynamic as it gets different set of weights at different parts of the data.

The method does not need a second level of model. It defines a neighbourhood and checks number of correct classifications done by different base learners. The weights provided to different base learners come from the number of correct classifications done by those base learners in the neighbourhood of the point to be classified. The detailed calculations can be understood in the Sect. 4.1.

4.1 Steps of Stacking Using Neighbourhood-Based Dynamic Weights

Step 1 Run k number of classification techniques $T1, T2, \ldots, Tk$ and build models.

Step 2 Get the first observation $O1$ from the new dataset.

Step 3 Using the chosen distance measure—(say, Euclidean if the independent variables are continuous) find out distance of all of the points in the training dataset from $O1$.

Step 4 Choose 'n' nearest neighbours of $O1$ using the distances found.

Step 5 Check correct classifications done by different base learners in the neighbourhood of $O1$.

Step 6 Derive the weight of the base learner Ti [$i = 1, 2, 3, \ldots, k$] as the ratio of 'correct classifications done by Ti in the neighbourhood of $O1$' to 'sum of correct classifications done by all of the techniques' in the neighbourhood of O1. Say, the weight for base learner Ti [$i = 1, 2, 3, \ldots, k$] in the neighbourhood of $O1$ is $w_{1i}[i = 1, 2, 3, \ldots, k]$.

Step 7 Do a dot-product of the weights of the base learners with the probabilities of event they give for $O1$. Say, the probability of event given by base learner Ti [$i = 1, 2, 3, \ldots, k$] for $O1$ is $p_{1i}[i = 1, 2, 3, \ldots, k]$. So, we have to get

$$\sum_{i=1}^{k} w_{1i} p_{1i}$$

Step 8 The number obtained at Step 7 can be considered as the probability of event for $O1$. Thereafter, classification is done based on a cut-off.

Step 9 Repeat same procedure for other observations from the new dataset like $O2$, $O3$, etc. The neighbourhood for different observations can be different, and hence, the weights they get for different base learners would be different.

Step 10 Hyperparameter tuning: Run the whole procedure for different values of n to get optimal value of n. Please note that n is the number of nearest neighbours as mentioned in Step 4.

5 Proposed Method II: Stacking Using Distance-Based Dynamic Weights

This method is a variant of the method described in proposed method I. Here, all points in the neighbourhood do not get same importance. Within the neighbourhood, distance of a point from the new observation, is considered to provide importance to the point while calculating weights of the base learners. Adjustment factors are calculated which are higher for a closer point to the new observation in comparison to a distant point. These adjustment factors are used to get an adjusted count of correct classifications by different base learners. The weights provided to different base learners come from the number of adjusted correct classifications done by those base learners in the neighbourhood of the new observation to be classified. The detailed calculations can be understood in Sect. 5.1.

5.1 Steps of Stacking Using Distance-Based Dynamic Weights

Step 1. Run k number of classification techniques $T1, T2, \ldots, Tk$ and build models.
Step 2. Get the first observation $O1$ from the new dataset.
Step 3. Using the chosen distance measure, find out distance of all of the points in the training dataset from $O1$.
Step 4. Choose 'n' nearest neighbours of $O1$ using the distances found.
Step 5. Find out the maximum distance a point has with $O1$ within O1s defined neighbourhood. Say the mentioned maximum distance is M.
Step 6. Get vector of adjustment factors **A** consisting of adjustment factors Aj [j = 1, 2, 3, …, n] for each of the n points as
1—(distance of the point from $O1$/ M)
The point with distance M from $O1$ will have 0 as adjustment factor. So, one of the n values in the vector **A** will be 0.
Step 7. Define Cij as 1 if technique Ti has done correct classification of the observation j in the defined neighbourhood of $O1$; otherwise, Cij is 0. [i = 1, 2, 3, …, k; j = 1, 2, 3, …, n].
Get matrix **C** of order $k \times n$.
Step 8. Get the matrix $\mathbf{U} = \mathbf{C} \cdot \mathbf{A}$. **U** is a matrix of order $k \times 1$
Step 9. Get the matrix **W** = **U**/sum of all elements of **U**. **W** is a matrix of order $k \times 1$. The elements of **W** in serial are weights to be used on the k base learners, respectively.
Step 10. Do a dot-product of the weights of the base learners with the probabilities of event they give for $O1$
Step 11. The number obtained at Step 10 can be considered as the probability of event for $O1$. Thereafter, classification is done based on a cut-off.

Step 12. Repeat same procedure for other observations from the new dataset like $O2$, $O3$, etc. The neighbourhood for different observations can be different, and hence, the weights they get for different base learners would be different.

Step 13. Hyperparameter tuning: Run the whole procedure for different values of n to get optimal value of n.

6 Findings

Four classification techniques, namely logistic regression, linear discriminant analysis, decision tree and k-nearest neighbors are run on four datasets freely available. Thereafter, we ran stacking by conventional way as well as both of the proposed ways where weights vary. Performance of the models are judged by the following metrices.

AUC: Area under the receiver operating characteristics (ROC) curve.

Accuracy: Percentage of correct classifications.

Precision: Percentage of truly being '1' (event) out of the total number of '1' predicted. Suppose, a model has classified $m1$ number of observations to the class of '1' while only m out of m1 are correctly classified as '1'. Then, precision of the model equals to $m/m1$.

Recall: Percentage of observed '1' (event) predicted as '1'. Suppose, the number of observations belonging to class '1' in the data is $m2$ out of which m are correctly classified as '1' by a model. Then, recall of the model equals to $m/m2$.

The datasets are split randomly in 80:20 ratio where distribution of the categories of the target variable are kept same. Models are built on the 80% of the dataset, and the remaining 20% is treated as if it is the set of new observations where the models are to be applied. Performance of models on 20% of the dataset is of more importance and are compared. Henceforth, the 80% chunk is referred as training data and the 20% hold-out sample is referred to as test data.

The findings based on the experimentation done on four datasets are given below. The probability cut-off for classifying event or non-event is set at the proportion of event in the whole dataset. The proposed methods of stacking are done using different values of 'n' which is the number of neighbours, and the optimum value on 'n' is chosen based on accuracy.

6.1 Wholesale Customer Data

The data is of clients of a wholesale distributor [3]. It is of 440 observations. The data has information about how much clients spent in a year on different categories, namely fresh products, milk products, grocery products, frozen products, paper products and delicatessen products. Annual spent on different categories are continuous variables. Channel is the target variable, which is a binary nominal variable having

Table 1 Evaluation of the results on the training dataset of wholesale customer data

Techniques	Accuracy (%)	Precision (%)	Recall (%)
Logistic regression	90	82	90
Linear discriminant analysis	85	94	58
Decision tree	94	93	88
K-nearest neighbor ($K = 15$)	92	87	89
Stacking by conventional way	94	93	89

Chosen value of K is the one where highest accuracy obtained when different values of K are tried

Table 2 Evaluation of the results on the test dataset of wholesale customer data

Techniques	AUC	Accuracy (%)	Precision (%)	Recall (%)
Logistic regression	0.97	92	84	93
Linear discriminant analysis	0.97	88	95	64
Decision tree	0.94	92	84	93
K-nearest neighbor ($K = 15$)	0.97	92	84	93
Stacking by conventional way	0.96	92	84	93

categories 'Horeca (hotel/restaurant/cafe)' and 'Retail'. There is one more variable which is on regions; we have not used it.

We have used different techniques aiming classification of the categories of the target variable 'Channel'. The models provide probability of being 'Retail' (Tables 1 and 2).

Though 'precision' is found to be better in training dataset in most of the techniques, 'recall' is found to be better in test data due to increase in correct classification of the events '1'. The only technique which did not perform at par with others is 'linear discriminant analysis'. We applied the new ways of stacking, accuracy, precision and recall on the test data which are 93%, 84% and 96%, respectively, when stacking is done using neighbourhood-based dynamic weights with size of neighbourhood $n = 5$ (found optimum among the values of $n = 2, 3, 4, \ldots, 25$). Stacking using distance-based dynamic weights with size of neighbourhood $n = 6$ (found optimum among the values of $n = 2, 3, 4, \ldots, 25$) also provide accuracy, precision and recall of 93%, 84% and 96%, respectively, on the test data. So, both of the new methods of stacking increases recall; as recall becomes 96% from 93%.

6.2 *Pima Indians Diabetes Data*

The data is of female patients of at least 21 years old of Pima Indian heritage [4]. It is of 768 observations. The data has information of patients about number of pregnancies, plasma glucose concentration at 2 h in an oral glucose tolerance test,

Table 3 Evaluation of the results on the training dataset of Pima Indians diabetes data

Techniques	Accuracy (%)	Precision (%)	Recall (%)
Logistic regression	74	60	80
Linear discriminant analysis	77	73	56
Decision tree	85	84	71
K-nearest neighbor ($K = 17$)	79	76	58
Stacking by conventional way	85	80	75

Chosen value of K is the one where highest accuracy obtained when different values of K are tried

Table 4 Evaluation of the results on the test dataset of Pima Indians diabetes data

Techniques	AUC	Accuracy (%)	Precision (%)	Recall (%)
Logistic regression	0.87	77	61	91
Linear discriminant analysis	0.87	77	69	65
Decision tree	0.77	75	65	59
K-nearest neighbor ($K = 17$)	0.80	72	63	50
Stacking by conventional way	0.81	75	65	65

diastolic blood pressure, triceps skinfold thickness (mm), 2 h serum insulin (mu U/ml), body mass index (weight in kg/(height in m)^2), diabetes pedigree function and age (years). Diabetes (Class) is the target variable, which is a binary nominal variable having categories 'diabetic' and 'non-diabetic'.

We have used different techniques aiming classification of the categories of the target variable 'diabetes (class)'. The models provide probability of being 'diabetic' (Tables 3 and 4).

Here, logistic regression performs much better than other techniques. Surprisingly, performance of logistic regression is much better in the test data set than training dataset. In case of linear discriminant analysis, its precision is 8% higher than the precision of logistic regression, but recall is 26% lower than recall of logistic regression. Performance of other techniques are not satisfactory, and thus, conventional stacking does not cause any benefit to the results when compared with logistic regression. However, both of the new methods of stacking does much better than the conventional way of stacking in recall though some compromise is there in precision. Accuracy, precision and recall on the test dataset in case of stacking using neighbourhood-based dynamic weights with size of neighbourhood $n = 4$ (found optimum among the values of $n = 2, 3, 4, \ldots, 25$) are 74%, 60%, 81%, respectively, while such measures on the test dataset when stacking using distance-based dynamic weights with size of neighbourhood $n = 6$ (found optimum among the values of $n = 2, 3, 4, \ldots, 25$) is used are 76%, 61%, 87%, respectively. So, we see that stacking using distance-based dynamic weights performed better than stacking using neighbourhood-based dynamic weights where accuracy, precision and recall got increased by 2%, 1% and 6%, respectively.

Table 5 Evaluation of the results on the training dataset of bank note authentication data

Techniques	Accuracy (%)	Precision (%)	Recall (%)
Logistic regression	99	98	99
Linear discriminant analysis	97	95	100
Decision tree	98	97	99
K-nearest neighbor ($K = 9$)	100	100	100
Stacking by conventional way	100	100	100

Chosen value of K is the one where highest accuracy obtained when different values of K are tried

Table 6 Evaluation of the results on the test dataset of bank note authentication data

Techniques	AUC	Accuracy (%)	Precision (%)	Recall (%)
Logistic regression	0.99	99	98	99
Linear discriminant analysis	0.99	99	97	100
Decision tree	0.98	99	98	99
K-nearest neighbor ($K = 9$)	1.00	100	100	100
Stacking by conventional way	1.00	100	100	100

6.3 Bank Note Authentication Data

The data is of authentication of bank notes [5]. It is of 1372 observations. Data were extracted from images that were taken from genuine and forged banknote-like specimens. Wavelet transform tool was used to extract features from images. The data has information about variance of wavelet transformed image, skewness of wavelet transformed image, kurtosis of wavelet transformed image, entropy of image which are continuous variables. Class is the target variable, which is a binary nominal variable having categories 'authenticate' or 'non-authenticate'.

We have used different techniques aiming classification of the categories of the target variable 'Class'. The models provide probability of being 'authenticate' (Tables 5 and 6).

Here, recall of all of the base techniques is between 99% and 100%. Here, all of the methods of stacking provide 100% in all three measures, namely accuracy, precision and recall. The size of neighbourhood n used for both of the proposed methods is 3.

6.4 Iris Data

The data is of different types of flowers [6]. It is of 150 observations. The data has information about sepal length, sepal width, petal length, petal width and species type. Species is the categorical variable, having categories 'setosa', 'versicolor' and

Table 7 Evaluation of the results on the training dataset of iris data

Techniques	Accuracy (%)	Precision (%)	Recall (%)
Logistic regression	73	58	75
Linear discriminant analysis	73	63	43
Decision tree	95	89	98
K-nearest neighbor ($K = 11$)	96	93	95
Stacking by conventional way	96	93	95

Chosen value of K is the one where highest accuracy obtained when different values of K are tried

Table 8 Evaluation of the results on the test dataset of iris data

Techniques	AUC	Accuracy (%)	Precision (%)	Recall (%)
Logistic regression	0.94	87	80	80
Linear discriminant analysis	0.95	80	83	50
Decision tree	1	100	100	100
K-nearest neighbor ($K = 11$)	1	100	100	100
Stacking by conventional way	1	100	100	100

'virginica'; other variables are continuous in nature which are used as explanatory variables. Spec_1 is our derived variable used as target variable which is binary in nature stating whether the species is 'versicolor' or not.

We have used different techniques aiming classification of the categories of the target variable 'Spec_1'. The models provide probability of being 'versicolor' (Tables 7 and 8).

Here, all of the methods of stacking provide 100% in all three measures, namely accuracy, precision and recall. The size of neighbourhood n used for both of the proposed methods is 3.

7 Conclusion

Both of the proposed methods of stacking with dynamic weights work better than the conventional way of stacking since the proposed methods are flexible. The performance of stacking by conventional way and the proposed methods is shown below (Table 9).

Table 9 Performance comparison

Dataset	Method	Accuracy (%)	Precision (%)	Recall (%)
Wholesale Customer Data	Stacking by conventional way	92	84	93
	Stacking by neighbourhood-based dynamic weights	93	84	96
	Stacking by distance-based dynamic weights	93	84	96
Pima Indians Diabetes Data	Stacking by conventional way	75	65	65
	Stacking by neighbourhood-based dynamic weights	74	60	81
	Stacking by distance-based dynamic weights	76	61	87
Bank Note Authentication Data	Stacking by conventional way	100	100	100
	Stacking by neighbourhood-based dynamic weights	100	100	100
	Stacking by distance-based dynamic weights	100	100	100
Iris Data	Stacking by conventional way	100	100	100
	Stacking by neighbourhood-based dynamic weights	100	100	100
	Stacking by distance-based dynamic weights	100	100	100

Though 'accuracy' and 'precision' are not seen to get improved by the proposed methods of stacking, but improvement in 'recall' is noticed. In the 'wholesale customer data', recall is as high as 93% by conventional stacking. Still by applying both of the proposed methods, we are able to increase recall by further 3%. In the Pima Indian Diabetes Data, stacking by neighbourhood-based dynamic weights has performed poorer in precision by 5% but has increased recall by 16% over stacking by conventional way. Results of stacking using distance-based dynamic weights is even more encouraging as we get 22% increase in recall over stacking by conventional way, and a compromise of 4% is there in precision. Performances of the proposed methods of stacking are same as the one by conventional way on other two datasets as the conventional way did not leave any scope of improvement here by hitting 100% in all of the measures.

Hence, both of the proposed methods 'stacking by neighbourhood-based dynamic weights' and 'stacking by distance-based dynamic weights' are seen to outperform conventional way of stacking in recall.

Acknowledgements We hereby feel happy to acknowledge Siddhartha Mukherjee of Tata Consultancy Services, Kolkata and thank him for his help in coding.

References

1. Wolpert, D. (1992). Stacked generalization. *Neural Networks, 5*(2), 241–259.
2. Breiman, L. (1996). Stacked regression. *Machine Learning, 24.*
3. Cardoso, M. G. M. S. (2014). [Wholesale Customer Data Set] UCI Machine Learning Repository [https://archive.ics.uci.edu/ml]. Irvine, CA: University of California, School of Information and Computer Science.
4. Sigillito, V. (1990). [Pima Indians Diabetes Data Set] UCI Machine Learning Repository [https://archive.ics.uci.edu/ml]. Irvine, CA: University of California, School of Information and Computer Science.
5. Lohweg, V., & Doerksen, H. (2013). [Banknote Authentication Data Set] UCI Machine Learning Repository [https://archive.ics.uci.edu/ml]. Irvine, CA: University of California, School of Information and Computer Science.
6. Fisher, R. A., & Marshall, M. (1988). [Iris Data Set] UCI Machine Learning Repository [https://archive.ics.uci.edu/ml]. Irvine, CA: University of California, School of Information and Computer Science.

The Effect of Infrastructure and Taxation on Economic Growth: Insights from Middle-Income Countries

Rudra P. Pradhan

1 Introduction

The connotation between infrastructure and economic growth has been the subject of considerable academic research over the past couple of decades (Holmgren and Merkel 2017; Pradhan et al. 2018a, b; WDR 1994). Various studies have concentrated on diverse countries, time periods, statistical techniques and altered proxy variables which have been used for reviewing the infrastructure-growth relationship (Pradhan et al. 2015, 2016; Canning 1999; Duggal et al. 1999; Holtz-Eakin and Schwartz 1995). From the available studies, we observe that the nexus between infrastructure and economic growth is rather indecisive and there is a consent neither on the existence nor on the direction of causality. A foremost reason for the absence of consensus is that the Granger causality test in a bivariate framework is likely to be biased due to the omission of relevant variables affecting infrastructure and economic growth nexus (Pradhan et al. 2019; Stoilova 2017; Besley and Persson 2013). This calls for studying the relationship between the two using a multivariate framework. No doubt there are couple of existing studies that examine the relationship between the two using a multivariate framework by incorporating different macroeconomic variables that affect the infrastructure-growth linkage (Pradhan et al. 2017, 2018c; Barro 1991). In this paper, we intend to examine the relationship between infrastructure and economic growth by incorporating taxation into the system. There are many different ways we can justify the inclusion and importance of taxation to the infrastructure-growth nexus (see, inter alia, Pradhan et al. 2020; Chauvet and Ferry 2016; Yoshino and Adidhadjaev 2016; Besley and Persson 2013). The remainder of this paper is organized as follows. Section 2 describes our model and data. Section 3 describes the results. Section 4 offers conclusion.

R. P. Pradhan (✉)
Indian Institute of Technology Kharagpur, Kharagpur, India
e-mail: rudrap@vgsom.iitkgp.ernet.in

A. K. Laha (ed.), *Applied Advanced Analytics*, Springer Proceedings in Business and Economics, https://doi.org/10.1007/978-981-33-6656-5_16

2 Methods of Study and Data

The paper uses the following panel data modelling to investigate the effect of infrastructure and taxation on economic growth.

The below regression model is set for this investigation.

$$\text{Economic Growth} = f(INF, TAX, Z) \tag{1}$$

$$\text{PEG}_{it} = \alpha\text{PEG}_{it-1} + \beta\text{INF}_{it} + \delta\text{TAX}_{it} + \mu Z_{it} + \theta_i + \varepsilon_{it} \tag{2}$$

where $\varepsilon_{it} = \eta_i + \nu_t + \zeta_{it}$; η_i is country effect; ν_t is time effect and ζ_{it} is independent and identically distributed among countries (i) and years (t).

PEG is the growth rate of real per capita GDP; INF is infrastructure,[1] TAX is the tax propensity,[2] Z is set of macroeconomic control variables, which includes government expenditure [GOE], gross capital formation [GCF], foreign direct investment [FDI], inflation [INF], human capital [HUC], and population growth [POG]. We expect that infrastructure and tax have positive impact on economic growth.

The study uses a panel dataset covering the selected middle-income countries (MICs) for the period 1970–2016. The countries include in the analysis are listed in Table 4 (see Appendix 1).

The data were obtained from *World Development Indicators* by the World Bank, Washington DC.

3 Empirical Results

The dynamic panel data model specified in Eq. (2) is positioned to estimate the impact of infrastructure and taxation on per capita economic growth. The estimated results are shown in Tables 1, 2 and 3, depending upon the use of three subsets. First subset is for upper middle-income countries (UMICs), second subset is for lower middle-income countries (LMICs), and third subset is for total middle-income countries (MICs), combined UMICs and LMICs. Each subset has six different cases, depending upon the use of six ICT infrastructure indicators, namely TLL, MOB, INU, INS, FIB, and ICI. These are as follows:

Case I [C1] deals with the relationship between PEG_t, PEG_{t-1}, TAX_t, TRA_t, TLL_t, GEX_t, GCF_t, FDI_t, OPE_t, INF_t, POG_t, and HCI_t.

---

[1]It includes both ICT infrastructure and transportation infrastructure. ICT infrastructure includes the use of telephone land lines (TLL), mobile phones (MOB), Internet users (INU), Internet servers (INS), fixed broadband (FIB), and a composite index (ICI), while transportation infrastructure includes the use of road length, and railways length. Principal component analysis (PCA) is deployed to derive the ICI from the other five ICT infrastructure indicators. Detailed derivation is available from the authors upon request.

[2]It is used as tax revenue as a percentage of gross domestic product.

Table 1 Dynamic panel data estimates (UMICs)

Variables	C1	C2	C3	C4	C5	C6
PEG_{t-1}	0.09*	0.10*	0.09*	0.10*	0.10*	0.09*
TAX	−0.07	−0.06	−0.06	−0.08	−0.01	−0.04
TRA	0.02*	0.04*	0.04*	0.04*	0.04*	0.04*
ICT	0.05*	0.01*	0.01*	0.02	0.04*	0.01
GEX	−0.18*	−0.18*	−0.18*	−0.18*	−0.19*	−0.17*
GCF	0.18*	0.20*	0.19*	0.20*	0.21*	0.21*
FDI	0.14*	0.13**	0.12*	0.10**	0.10*	0.12*
OPE	0.03	0.02	−0.01	0.02	0.02	0.03
INF	−0.08*	−0.08*	−0.08*	−0.10*	−0.10*	−0.08*
POG	−0.06***	−0.04*	−0.02	−0.01	−0.01	−0.04
HCI	0.03	0.01	0.02	0.01	0.01	0.01
Constant	1.12*	1.13*	1.15*	1.14*	1.14*	1.14*
Wald X^2	180.7*	177.6*	178.5*	169.5*	169.7*	111.4*

Note 1: PEG_{t-1} is lagged per capita economic growth, TAX is tax revenue, TRA is transportation infrastructure, ICT is ICT infrastructure, GEX is government expenditure, GCF is gross capital formation, FDI is foreign direct investment, OPE is trade openness, INF is inflation rate, POG is population growth, HCI is human capital index, and UMICs is upper middle-income countries
Note 2: PEG is dependent variable and indicates per capita economic growth
Note 3: C1–C6 are different cases, depending upon the inclusion of six ICT indicators, namely telephone land lines, mobile phones, Internet users, Internet servers, fixed broadband, and a composite index of ICT infrastructure (ICI). These cases are well-defined in the text
Note 4: *, **, ***: Indicates statistical level of significance at 1–10% levels

Case 2 [C2] deals with the relationship between PEG_t, PEG_{t-1}, TAX_t, TRA_t, MOB_t, GEX_t, GCF_t, FDI_t, OPE_t, INF_t, POG_t, and HCI_t.

Case 3 [C3] deals with the relationship between PEG_t, PEG_{t-1}, TAX_t, TRA_t, INU_t, GEX_t, GCF_t, FDI_t, OPE_t, INF_t, POG_t, and HCI_t.

Case 4 [C4] deals with the relationship between PEG_t, PEG_{t-1}, TAX_t, TRA_t, INS_t, GEX_t, GCF_t, FDI_t, OPE_t, INF_t, POG_t, and HCI_t.

Case 5 [C5] deals with the relationship between PEG_t, PEG_{t-1}, TAX_t, TRA_t, FIB_t, GEX_t, GCF_t, FDI_t, OPE_t, INF_t, POG_t, and HCI_t.

Case 6 [C6] deals with the relationship between PEG_t, PEG_{t-1}, TAX_t, TRA_t, ICI_t, GEX_t, GCF_t, FDI_t, OPE_t, INF_t, POG_t, and HCI_t.

The estimated coefficients of ICT indicators indicate that they are positively associated with per capita economic growth in all the three subsets. However, the coefficients of ICT indicators are statistically significant in few occasions and varies from case to case (see Tables 1, 2, 3 and cases C1–C6 in each subset). They estimated coefficients range from 0.002 to 0.022 basis point. The point estimate implies that a 100% increase in ICT infrastructure (for TLL, MOB, INU, INS, FIB, and ICI) is associated with 2–22 basis point increase in per capita economic growth.

Table 2 Dynamic panel data estimates (LMICs)

Variables	C1	C2	C3	C4	C5	C6
PEG_{t-1}	0.20*	0.203*	0.203*	0.203*	0.203*	0.203*
TAX	−0.02*	−0.017*	−0.017*	−0.017*	−0.017*	−0.018*
TRA	0.007*	0.007*	0.007*	0.007*	0.007*	0.007*
ICT	0.027*	0.004	0.001	0.002	0.002	0.002
GEX	−0.085*	−0.292*	−0.293*	−0.291*	−0.292*	−0.288*
GCF	0.083*	0.081*	0.081*	0.081*	0.081*	0.081*
FDI	0.015	0.009	0.007	0.009	0.008	0.011
OPE	0.152*	0.142*	0.141*	0.141*	0.143*	0.141*
INF	−0.108*	−0.104*	−0.103*	−0.103*	−0.104*	−0.103*
POG	−0.164*	−0.109*	−0.106*	−0.110*	−0.108*	−0.115*
HCI	0.001	0.001	0.001	0.001	0.001	0.001*
Constant	1.176*	1.178*	1.179*	1.179*	1.178*	1.175*
Wald X^2	122.1*	110.5*	110.9*	111.4*	110.6*	111.4*

Note 1: PEG_{t-1} is lagged per capita economic growth, TAX is tax revenue, TRA is transportation infrastructure, ICT is ICT infrastructure, GEX is government expenditure, GCF is gross capital formation, FDI is foreign direct investment, OPE is trade openness, INF is inflation rate, POG is population growth, HCI is human capital index, and LMICs is lower middle-income countries
Note 2: PEG is dependent variable and indicates per capita economic growth
Note 3: C1–C6 are different cases, depending upon the inclusion of six ICT indicators, namely telephone land lines, mobile phones, Internet users, Internet servers, fixed broadband, and a composite index of ICT infrastructure (ICI). These cases are well-defined in the text
Note 4: *, **, ***: Indicates statistical level of significance at 1–10% levels

The estimated coefficients of transportation infrastructure indicate that it is positively related with current per capita economic growth. They estimated coefficients range from 0.007 to 0.030 basis point, and they are statistically significant at 1% level in all the three cases. The point estimate implies that a 10% increase in transportation is associated with 0.7–3.0 basis point increase in current year per capita economic growth.

The estimated coefficients of lagged economic growth indicate that it is positively related with current per capita economic growth. They estimated coefficients range from 0.015 to 0.203 basis point, and they are statistically significant at 1% level in all the three cases. The point estimate implies that as economic growth itself is a derived quantity of 10% increase in last year is associated with 1.5–20.3 basis point increase in current year per capita economic growth.

The estimated coefficients of gross capital formation indicate that it is positively related with per capita economic growth. They estimated coefficients range from 0.036 to 0.206 basis point, and they are statistically significant at 1% level in all the three cases. The point estimate implies that a 10% increase in gross capital formation is associated with 0.36–2.036 basis point increase in per capita economic growth.

Table 3 Dynamic panel data estimates (MICs)

Variables	C1	C2	C3	C4	C5	C6
PEG_{t-1}	0.053*	0.015*	0.015*	0.015*	0.015*	0.015*
TAX	−0.089	−0.018	−0.002	−0.012	−0.002	−0.002
TRA	0.030*	0.030*	0.030*	0.03*	0.030*	0.030*
ICT	0.022*	0.004*	0.001	0.003	0.002	0.002
GEX	−0.066*	−0.063*	−0.063*	−0.067*	−0.064*	−0.067*
GCF	0.036*	0.038*	0.037*	0.037*	0.037*	0.038*
FDI	0.060**	0.072*	0.074*	0.072*	0.073*	0.069*
OPE	0.022	0.030*	0.031*	0.032	0.031	0.03**
INF	−0.408*	−0.041*	−0.042*	−0.041*	−0.042*	−0.041*
POG	−0.187*	−0.014*	−0.013*	−0.145	−0.138*	−0.149*
HCI	0.005	0.007*	0.007*	0.001*	0.001*	0.001*
Constant	1.473*	1.448*	1.458*	1.446*	1.448*	1.448*
Wald X^2	364.7*	360.9*	360.8*	360.9*	361.0*	361.1*

Note 1: PEG_{t-1} is lagged per capita economic growth, TAX is tax revenue, TRA is transportation infrastructure, ICT is ICT infrastructure, GEX is government expenditure, GCF is gross capital formation, FDI is foreign direct investment, OPE is trade openness, INF is inflation rate, POG is population growth, HCI is human capital index, and UMICs is upper middle-income countries
Note 2: PEG is dependent variable and indicates per capita economic growth
Note 3: C1–C6 are different cases, depending upon the inclusion of six ICT indicators, namely telephone land lines, mobile phones, Internet users, Internet servers, fixed broadband, and a composite index of ICT infrastructure (ICI). These cases are well-defined in the text
Note 4: *, **, ***: Indicates statistical level of significance at 1–10% levels

The estimated coefficients of foreign direct investment indicate that it is positively related with per capita economic growth. They estimated coefficients range from 0.007 to 0.142 basis point, and they are statistically significant at 1% level in all the three cases. The point estimate implies that a 100% increase in foreign direct investment is associated with 0.70–14.2 basis point increase in per capita economic growth.

The estimated coefficients of income tax rate indicate that it is negatively linked with per capita economic growth. They range from −0.017 to −0.018 basis point and are statistically significant at 10% levels of significance in middle-income countries. The point estimate implies that a 100% increase in tax propensity is associated with 1.7–1.8 basis point decrease in per capita economic growth.

The estimated coefficients of government expenditure indicate that it is negatively linked with per capita economic growth. They range from −0.063 to −0.293 basis point and are statistically significant at 10% levels of significance in all the three subsets. The point estimate implies that a 100% increase in government expenditure is associated with 6.3–29.3 basis point decrease in per capita economic growth. This might be the possibility of the occurrence of unproductive expenditure.

Table 4 List of MICs

Country name	Region	Income group
Albania	South Eastern Europe	UMICs
Algeria	Middle East and North Africa	LMICs
American Samoa	South Central Pacific Ocean	UMICs
Angola	Sub-Saharan Africa	LMICs
Armenia	Central Africa	LMICs
Azerbaijan	Western Asia and Eastern Europe	UMICs
Bangladesh	South Asia	LMICs
Belarus	Eastern Europe	UMICs
Belize	Latin America and Caribbean	UMICs
Bhutan	South Asia	LMICs
Bolivia	Western-Central South America	LMICs
Bosnia and Herzegovina	South East Europe	UMICs
Botswana	South Eastern Africa	UMICs
Brazil	South America	UMICs
Bulgaria	South East Europe	UMICs
Cabo Verde	Sub-Saharan Africa	LMICs
Cambodia	Sub-Saharan Africa	LMICs
Cameron	Sub-Saharan Africa	LMICs
China	East Asia	UMICs
Colombia	Latin America and Caribbean	UMICs
Congo Republic	Sub-Saharan Africa	LMICs
Costa Rica	North America	UMICs
Cote d'Ivoire	Sub-Saharan Africa	LMICs
Cuba	Latin America and Caribbean	UMICs
Djibouti	Middle East and North Africa	LMICs
Dominica	Latin America and Caribbean	UMICs
Dominican Republic	Latin America and Caribbean	UMICs
Ecuador	South America	UMICs

(continued)

Table 4 (continued)

Country name	Region	Income group
Egypt Arab Republic	Middle East and North Africa	LMICs
El Salvador	Latin America and Caribbean	LMICs
Equilateral Guinea	Central America	UMICs
Fiji	Pacific Ocean	UMICs
Gabon	Central Africa	UMICs
Georgia	Western Asia and Europe	LMICs
Ghana	Sub-Saharan Africa	LMICs
Grenada	Latin America and Caribbean	UMICs
Guatemala	Central America	LMICs
Guyana	Latin America and Caribbean	LMICs
Honduras	Latin America and Caribbean	LMICs
India	South Asia	LMICs
Indonesia	South East Asia	LMICs
Iran Islamic republic	West Asia	UMICs
Iraq	West Asia	UMICs
Jamaica	Latin America and Caribbean	UMICs
Jordan	South West Asia	UMICs
Kazakhstan	Central Asia	UMICs
Kenya	Sub-Saharan Africa	LMICs
Kiribati	East Asia and Pacific	LMICs
Kosovo	Europe	LMICs
Kyrgyz Republic	Europe and Central Asia	LMICs
Lao PDR	East Asia and Pacific	LMICs
Lebanon	Middle East	UMICs
Lesotho	Sub-Saharan Africa	LMICs
Libya	North Africa	UMICs
Macedonia	South East Europe	UMICs
Malaysia	South East Asia	UMICs
Maldives	South West Asia and the Middle East	UMICs
Marshall Islands	Central Pacific Ocean	UMICs

(continued)

Table 4 (continued)

Country name	Region	Income group
Mauritius	South East Africa	UMICs
Mexico	North America	UMICs
Montenegro	South East Europe	UMICs
Micronesia	Pacific Ocean	LMICs
Mauritania	Sub-Saharan Africa	LMICs
Micronesia	East Asia and Pacific	LMICs
Moldova	European and Central Asia	LMICs
Mongolia	East Asia and Pacific	LMICs
Morocco	Middle East and North Africa	LMICs
Myanmar	East Asia and Pacific	LMICs
Namibia	South East Asia	UMICs
Nauru	Pacific Ocean	LMICs
Nicaragua	Latin America and Caribbean	LMICs
Nigeria	Sub-Saharan Africa	LMICs
Pakistan	South Asia	LMICs
Papua New Guinea	East Asia and Pacific	LMICs
Paraguay	South America	UMICs
Peru	South America	UMICs
Philippines	East Asia and Pacific	LMICs
Romania	South Eastern Europe	UMICs
Russian Federation	Eastern Europe and Northern Asia	UMICs
Samoa	Africa	LMICs
Sao Tome and Principe	Sub-Saharan Africa	LMICs
Serbia	South East Europe	UMICs
Solomon Islands	Sub-Saharan Africa	LMICs
Sudan	Central Africa	LMICs
South Africa	Africa	UMICs
St' Lucia	Latin America and Caribbean	UMICs
St. Vincent and the Grenadines	Eastern Caribbean	UMICs
Suriname	South America	UMICs
Sri Lanka	South Asia	LMICs
Swaziland	Southern Africa	LMICs

(continued)

Table 4 (continued)

Country name	Region	Income group
Thailand	South East Asia	UMICs
Timor-Leste	East Asia and Pacific	LMICs
Tonga	Pacific Ocean	UMICs
Tunisia	Middle East and North Africa	LMICs
Turkey	Western Asia	UMICs
Turkmenistan	Central Asia	UMICs
Tuvalu	Pacific Ocean	UMICs
Ukraine	European and Central Asia	LMICs
Uzbekistan	European and Central Asia	LMICs
Vanuatu	East Asia and Pacific	LMICs
Venezuela RB	South America	UMICs
Vietnam	East Asia and Pacific	LMICs
West Bank and Gaza	Middle East and North Africa	LMICs
Zambia	Sub-Saharan Africa	LMICs

Note: UMICs is upper middle-income countries; and LMICs is lower middle-middle inincome countries

The estimated coefficients of inflation rate indicate that it is negatively linked with per capita economic growth. They range from 0.041 to 0.108 basis point and are statistically significant at 1% level in all the three cases. This indicates that when the inflation rate increases by 100%, the per capita economic growth turned out to decrease by 4.1–10.8 basis point.

The estimated coefficients of population growth rate indicate that it is negatively linked with per capita economic growth. They range from 0.106 to 0.149 basis point and are statistically significant at 1% level in all the three cases. This indicates that when the population growth rate increases by 10%, the per capita economic growth turned out to decrease by 1.06–14.9 basis point.

The estimated coefficients of human capital indicate that it is positively linked with per capita economic growth. They range from 0.001 to 0.007 basis point and are statistically significant at 5% level in all the three cases. This indicates that when the unemployment rate increases by 100%, the per capita economic growth turns out to decrease by 0.01–0.007 basis point.

To sum up, the effect of infrastructure on per capita economic growth is positive and significant across the three subsets and in all six cases [C1–C6]. On the contrary, the effect of taxation on economic growth is negative and the impact varies from UMICs to LMICs. Furthermore, the regression coefficients of other macroeconomic variables are mostly consistent with the standard results in the existing literature. In some cases, the impact is positive on economic growth, while the impact is negative in other occasions. That means the findings are consistent with theoretical arguments and quite robust to different measures of ICT infrastructure including the country and year fixed effects.

4 Conclusion and Policy Implications

The paper started with two standard questions, "how does infrastructure affect economic growth?" and "how does tax propensity affect economic growth?" Our answer to this question is very much certain in the case of infrastructure and is true for both ICT infrastructure and transport infrastructure. The answer is also equally certain in the case of taxation, but it shows negative impact on economic growth.

Acknowledgements This paper has benefited from the helpful comments of the anonymous reviewers and Prof. Arnab K. Laha, the convenor of ICADABAI 2019, and the Editor of this volume, to whom we are grateful.

Appendix 1: List of Middle-Income Countries (MICs)

See Table 4.

References

Barro, R. (1991). Economic growth in a cross-section of countries. *Quarterly Journal of Economics, 106*(2), 407–444.

Besley, T., & Persson, T. (2013). Taxation and Development. Working paper. Accessed at https://econ.lse.ac.uk/staff/tbesley/papers/TaxationAndDevelopment.pdf.

Canning, D. (1999). *Infrastructure's contribution to aggregate output* (Vol. 2246). Washington DC: World Bank.

Chauvet, L., & Ferry, M. (2016). Taxation, infrastructure, and firm performance in developing countries. WIDER Working Paper, No. 2016/103. United Nations University, UNU-WIDER, Helsinki.

Duggal, V. G., Saltzman, C., & Klein, L. R. (1999). Infrastructure and productivity: A non-linear approach. *Journal of Econometrics, 92*(1), 47–74.

Holmgren, J., & Merkel, A. (2017). Much ado about nothing? A meta-analysis of the relationship between infrastructure and economic growth. *Research in Transportation Economics, 63,* 13–26.

Holtz-Eakin, D., & Schwartz, A. E. (1995). Infrastructure in a structural model of economic growth. *Regional Science and Urban Economics, 25*(2), 131–151.

Pradhan, R. P., Arvin, M. B., & Norman, N. R. (2015). The dynamics of information and communications technologies infrastructure, economic growth, and financial development: Evidence from Asian countries. *Technology in Society, 42*(1), 135–149.

Pradhan, R. P., Arvin, M. B., Mittal, J., & Bahmani, S. (2016). Relationships between telecommunications infrastructure, capital formation, and economic growth. *International Journal of Technology Management, 70*(2–3), 157–176.

Pradhan, R. P., Arvin, M. B., Nair, M., Mittal, J., & Norman, N. R. (2017). Telecommunications infrastructure and usage and the FDI–growth nexus: Evidence from Asian-21 countries. *Information Technology for Development, 23*(2), 235–260.

Pradhan, R. P., Arvin, M. B., Bahmani, S., Hall, J. H., & Bennett, S. E. (2018). Mobile telephony, economic growth, financial development, foreign direct investment, and imports of ICT goods: The case of the G-20 countries. *Journal of Industrial and Business Economics, 45*(2), 279–310.

Pradhan, R. P., Mallik, G., Bagchi, T. P., & Sharma, M. (2018). Information communication technology penetration and stock markets-growth nexus: From cross country panel evidence? *International Journal of Services Technology and Management, 24*(4), 307–337.

Pradhan, R. P., Mallik, G., & Bagchi, T. P. (2018). Information communication technology (ICT) infrastructure and economic growth: A causality evinced by cross-country panel data. *IIMB Review, 30,* 91–103.

Pradhan, R. P., Arvin, M. B., Nair, M., Bennett, S. E., & Hall, J. H. (2019). The information revolution, innovation diffusion and economic growth: An examination of causal links in european countries. *Quality and Quantity, 53,* 1529–1563.

Pradhan, R. P., Arvin, M. B., Nair, M., Bennett, S. E., & Bahmani, S. (2020). Some determinants and mechanics of economic growth in middle-income countries: The role of ICT infrastructure development, taxation and other macroeconomic variables. Singapore Economic Review (forthcoming).

Stoilova, D. (2017). Tax structure and economic growth: Evidence from the European Union. *Contaduria Administracion, 62,* 1041–1057.

WDR. (1994). *Infrastructure for development, world development report (WDR).* Washington DC: World Bank.

Yoshino, N., & Abidhadjaev, U. (2016). Impact of infrastructure investment on tax: Estimating spillover effects of the kyushu high-speed rail line in Japan on regional tax revenue. ADBI Working Paper Series, No. 574, Asian Development Bank Institute (ADBI), Tokyo.

Response Prediction and Ranking Models for Large-Scale Ecommerce Search

Seinjuti Chatterjee, Ravi Shankar Mishra, Sagar Raichandani, and Prasad Joshi

1 Problem Statement

User response prediction is the bread and butter of an ecommerce site. Every ecommerce site which is popular is running a response prediction engine behind the scenes to improve user engagement and to minimize the number of hops or queries that a user must fire in order to reach the destination item page which best matches the user's query. With the dawn of artificial intelligence (AI) and machine learning (ML), the whole merchandising process, web-commerce carousel product arrangement, personalized search results and user interactions can be driven by the click of a button. Modern-day ML platforms enable a dynamic cascade of models with different optimization functions which can be tuned towards a user's preference, taste and query trajectory.

In this paper, we talk about how Unbxd search services powers its user engagement and response prediction behind the scene using a plethora of optimized features across multiple channels and multiple domains. Elaborate feature engineering is deployed to understand the user's propensity to click. The search funnel lifecycle starts with a personalized search impression, captures a user click, progresses towards a cart and finally materializes into an order or sale. In this scenario, click through rate (CTR) modelling is the binary classification task of predicting whether a user would click given a ranked ordered set of products and conversion rate (CVR) modelling entails

S. Chatterjee (✉) · R. S. Mishra · S. Raichandani · P. Joshi
Unbxd, Bengaluru, India
e-mail: seinjuti@gmail.com

R. S. Mishra
e-mail: ravi@unbxd.com

S. Raichandani
e-mail: raichandanisagar@gmail.com

P. Joshi
e-mail: prasad.joshi@unbxd.com

A. K. Laha (ed.), *Applied Advanced Analytics*, Springer Proceedings in Business and Economics, https://doi.org/10.1007/978-981-33-6656-5_17

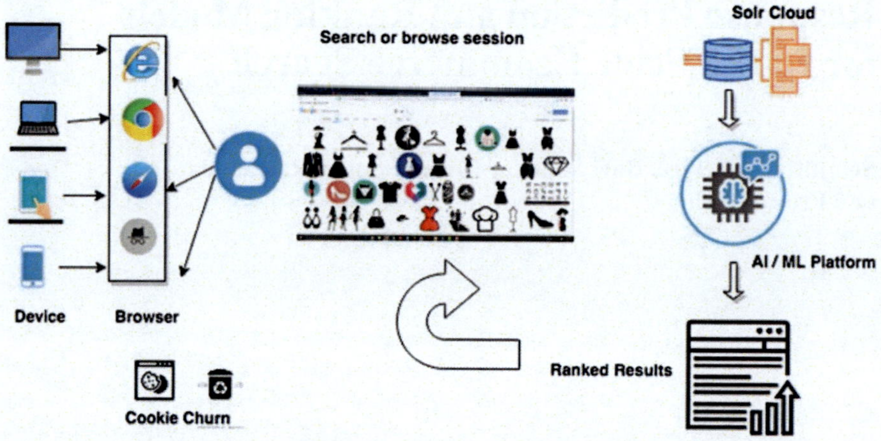

Fig. 1 View of user from search engine

the binary task of predicting whether user would purchase an item given he has shown click interest.

Figure 1 demonstrates the view of the user from a search engine's point and the view of the merchandizing website from the user's point. A user might be present in any device linked through the browser cookie or the device ID and might choose to initiate a search or browse session for the day; however, the click through and the conversion might potentially happen at a different device and at a different place (work or home) during a different time of day. Hence, an intelligent search engine must be able to stitch the user's trajectory seamlessly and understand the feature combinations leading up to an event for better engagement.

2 Literature Survey

As we started thinking about the bipartite query-product matching problem, we drew lot of similarity with the query-ads domain where personalizing the click prediction benefits both the users and the advertisers. In (Cheng & Cantu-Paz, 2010) the authors mention that the users will be presented ads in the manner that is most relevant to them, and the advertisers will receive clicks from users who are more engaged with the ads. In the search domain, however, most of the times search engines are burdened with the task of retrieval of the most relevant documents to improve precision and recall but in the process falling short of optimizing the business metrics like average order value, price per session, engagement rates such as click through, etc. In feature-based query performance as mentioned in (Kumar et al., 2018), the authors analyse user's behavioural patterns and build models to classify queries as high engagement queries, high sale through rate queries, thereby providing the search engine means to drive its business metrics other than just optimizing precision and recall per search. In (Zhou et al., 2018), researchers at Alibaba attempted to understand the deep interest graph of a user and the context of an ad, thereby using deep learning to model higher order feature interactions which drive a click. They have closely modelled the user's historical data as a sequence model and built a network which given the current sequence can closely predict the future interactions of the user in terms of product affinity and personalized ads. In (Guo et al., 2017), authors mention that during their study in a mainstream apps market, they found that people often download apps for food delivery at meal time, suggesting a second-order interaction between app category and time stamp.

In this paper, we talk about the search business insights that Unbxd has gathered being one of the largest ecommerce search service providers across domains like electronics, furniture, fashion, grocery. These insights indicate strong correlations between user, context, category, time of day features and the performance metrics of a query. Starting with the business problem, we have implemented distributed models at scale which now define our AI or ML framework. Together with our inhouse A/B testing framework, we have demonstrated the capability of our ML models to our clients, and the overall journey has been summarized in this paper.

3 Algorithm

The naive model we started with at Unbxd is a composition of clickability and buyability of a product learnt over historical clicks, carts, and orders. Here, we rank all products which are deemed relevant for a particular query in descending order of following score.

$$\text{Score} = \sum_{m=0}^{m=60}\left[\left(\text{clicks}/\left(\text{a}\,\text{e}^{b/r}\right)\right)(1+\ \text{carts/clicks})\left(1+\frac{\text{orders}}{\text{carts}}\right)^2\right]\text{e}^{-m/9} \tag{1}$$

where $r =$ rank of the product and a and b are constants and m is lookback days which is number of days in past we want to consider. This approach is a relatively static approach but captures the recency of clicks, carts, and orders and can be considered a ranking by popularity score. The composite score acts like an overall boost factor to be overlaid on indigenous search ranking implemented in Apache Solr Search Platform (Solr) in order to bubble up the trending products. However, this score is not nimble enough to adapt to dynamic ranking depending on the device or browser or query context or location or time of day.

Each impression consists of various attributes extracted from the request side parameters such as site, query, device, user, time of day, day of week, query category, location. This impression must be now matched with the Solr retrieved document attributes like product category, price, keywords, reviews, related products, tokens, etc. Hence, now the problem morphs into a bipartite graph matching algorithm with certain constraints and measured by the ranking loss function. Such features are called unigram features; since they only depend on one attribute; we can also use advanced features like query-dwell-time, time-to-first-click, time-to-first-cart, time-to-first-order, was-autosuggest-used, Wi-Fi-connection-type-of-user as mentioned in (Cheng & Cantu-Paz, 2010) depending on the data collection exposed through the search API.

Once we have collected such unigram features, we can fit the impressions data complete with the outcome to a **logistic regression** model as a binomial classification task which then estimates the probability of click given a new impression based on its features.

Logistic regression model: Given any input event, we assume that its outcome is a binary variable; i.e., it is either positive or negative. The logistic regression model calculates the probability of a positive outcome with the following function:

$$\text{logit}(p) = \log\left(\frac{p}{1-p}\right) \text{ where } p = 1/\left(1 + \mathrm{e}^{-w^T x}\right) \tag{2}$$

Here, x denotes vector of the features extracted from the input event and w is the vector of corresponding feature weight that we need to learn. LR model is trained with gradient descent. Assume that we have observed a set of events $X = \{x_i\}$ and their outcomes $Y = \{y_i\}$. Each event x_i can be represented with a set of features $\{x_{ij}\}$. We want to find the set of parameters w by maximizing the data likelihood $P(Y|X, w)$. This is equivalent to minimizing the following loss function:

$$L = -\log P(Y|X, w) = \sum_{i=0}^{n} \log P(y_i|x_i, w) \tag{3}$$

The beta coefficients of the model and ROC curve helps us understand the discriminating ability of the model between the positive and negative samples and ability to explain CTR through features.

4 Feature Selection

In Fig. 2, we show the factor map of the search session that is available to a third-party search engine. Some interesting features have been described below:

Data fields of a search session

- Outcome—click: 0/1 for non-click/click (can be cart or order depending on the model)
- Time Series Features
  - hour_of_day: int from 0 to 23, (parse format is YY-MM-DD-HH from session_time, so 14091123 means 23:00 on Sept. 11, 2014 UTC.)

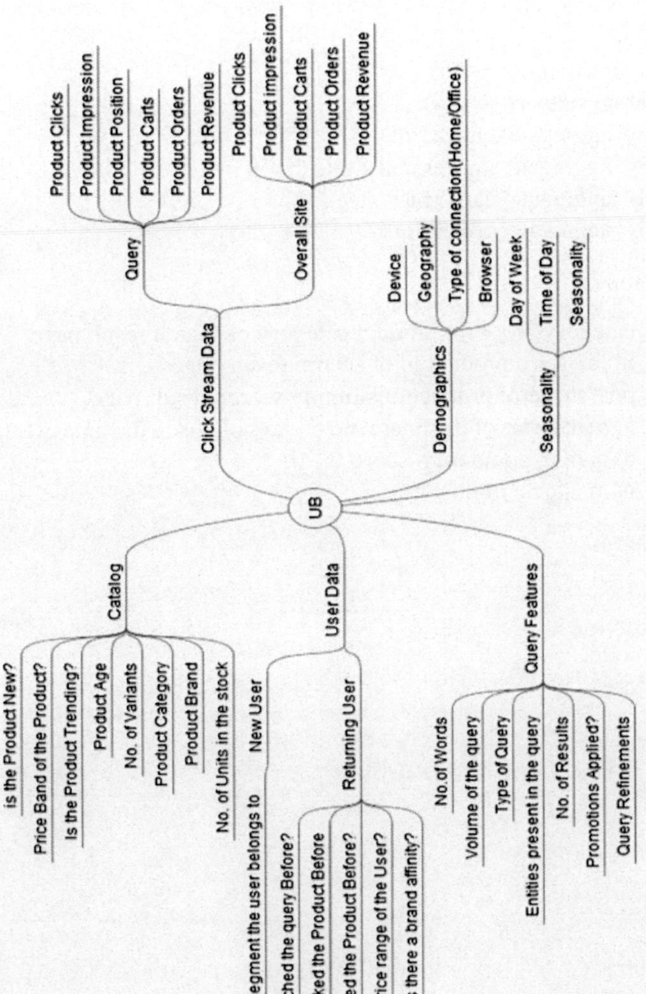

Fig. 2 Feature map of search session

  - day_of_week: 0–6 (parse format is YY-MM-DD-HH from session_time)
  - is_weekday: Boolean feature
  - is_weekend: Boolean feature
- Site Features
  - site_id: int
  - site_domain: string
  - site_catalog_size: int
  - site_daily_aggregate_clicks: int
  - site_daily_aggregate_impressions: int
  - site_daily_aggregate_carts: int
  - site_daily_aggregate_orders:int
- Product Features
  - product_category_in_serp: product category of search result page
  - product_id_in_serp: product id of search result page
  - product_pos_in_serp: product position in search result page
  - product_age (in terms of freshness: no of days old from the time of impression)
  - product_dynamic_popularity_score
  - product_num_stock_units
- Device Features
  - device_id
  - device_ip
  - device_model
  - device_type
- Location Features
  - country
  - pincode
  - region
  - latitude
  - longitude
- Query Features
  - query_tokens
  - query_length
  - query_refinements_in_same_session
  - query_entities (must have tokens, synonyms, entity recognition output—brand, product type, model no etc.)
  - query_daily_aggregate_clicks
  - query_daily_aggregate_orders
  - query_daily_aggregate_carts

- To account for position bias, we use a position-normalized statistic known as clicks over expected clicks (COEC) as defined in Cheng & Cantu-Paz, 2010

$$\text{COEC} = \frac{\sum_{r=0}^{R} c_r}{\sum_{r=0}^{R} ir_r * \text{CTR}_r}$$

where the numerator c_r is the total number of clicks received by a query-product pair; the denominator can be understood as the expected clicks (ECs) that an average product would receive after being impressed ir_r times at rank r, and CTR_r is the average CTR for each position in the result page (up to R), computed over all queries and products. We can obtain COEC statistic for specific query-product pairs, and this statistic is a good predictor of click probabilities as mentioned in (Cheng & Cantu-Paz, 2010). However, many data points are needed for this statistic to be significant but data for specific query-product pairs can be sparse and noisy.

In consideration of efficiency and robustness, we need to filter out certain types of features and this process is called **feature pruning**.

- Features with too few impressions. The simplest approach is to set a threshold on the number of impressions and filter out features with less impressions than the threshold. This step reduces the model size. However, one issue with this approach is that new features may get filtered out too easily. An alternative method is to put a threshold on the average number of impressions (averageImps) per feature defined as:

$$\text{average Imps} = \frac{\text{total Imps} + c}{\text{current Time} - \text{first Occurrence Time}} \tag{4}$$

where totalImps is total number of impressions received by the feature, currentTime is time of measurement in secs, firstOccurenceTime is time when feature received first impression and the constant c insures that new features will not be filtered out immediately.

- Features that are too old. If a feature is no longer active for a certain period, we may want to filter it out. In a fast-changing search space, features can become deprecated daily. Filtering such features out improves system efficiency.
- Features that have close to 0 weights. If a feature weight is close to 0, it means that this feature does not significantly affect the prediction, and if we filter this feature out, there should be little performance impact.

However, without expert feature engineering, exploring higher order features become daunting and learning sophisticated feature combinations behind user behavior is critical in maximizing CTR for search systems. The wide and deep model (Tze et al., 2016) from Google provides insights such as considering low- and high-order feature combinations simultaneously brings additional improvement over the cases of considering either alone. To this effect, we are in the process of implementing a factorization-machine-based neural network for cross product transformations of the original feature space, but it is work in progress.

5 Business Insights

We present in this section some of the business insights our analysts have come up with which provides the intuition behind feature-based response prediction.

Figure 3 provides the intuition that country, region, and zip code are differentiator signals for deciding the propensity of the user to purchase.
In Figs. 4 and 5, we show that by channel (mobile, desktop) and by day of week (weekday vs. weekend) our search sessions volumes and conversions vary. We see that the weekday traffic post 9 am comes mostly from desktop which indicates a user browsing or searching from workplace leading up to a lower average order value (AOV) compared to a user logging in the weekend over mobile when the AOV and engagement both peak, hence opening up an opportunity window for response prediction models to promote bigger ticket items for a query during this time and thereby maximizing conversions and AOV.

In the fourth and fifth graph Figs. 6 and 7, we compare new versus existing users and their search volumes and conversions over various channels—social media, email, display ads, private apps, organic search, etc. By tracking the user type and channel, response prediction models can effectively maximize CTR and CVR.

In the sixth and seventh graphs Figs. 8 and 9, we show how location signals and query category can be correlated. This opens up the opportunity to response predictor models to utilize the user's location (work or home) and the region to optimize the search results for certain query categories. However for staple products like laundry the business metrics remain fairly uniform as shown in Fig. 10.

In the last graph Fig. 11 we note that behaviour in cities is markedly different than behaviour in non-urban areas.

6 ML Architecture

Here, in Fig. 12 we present the details of the ML relevancy platform we have built at Unbxd and how we use the platform to power our scalable distributed logistic regression-based modelling workflow for response prediction.

Distributed LR Training Details in Spark

- The algorithm takes the following inputs:

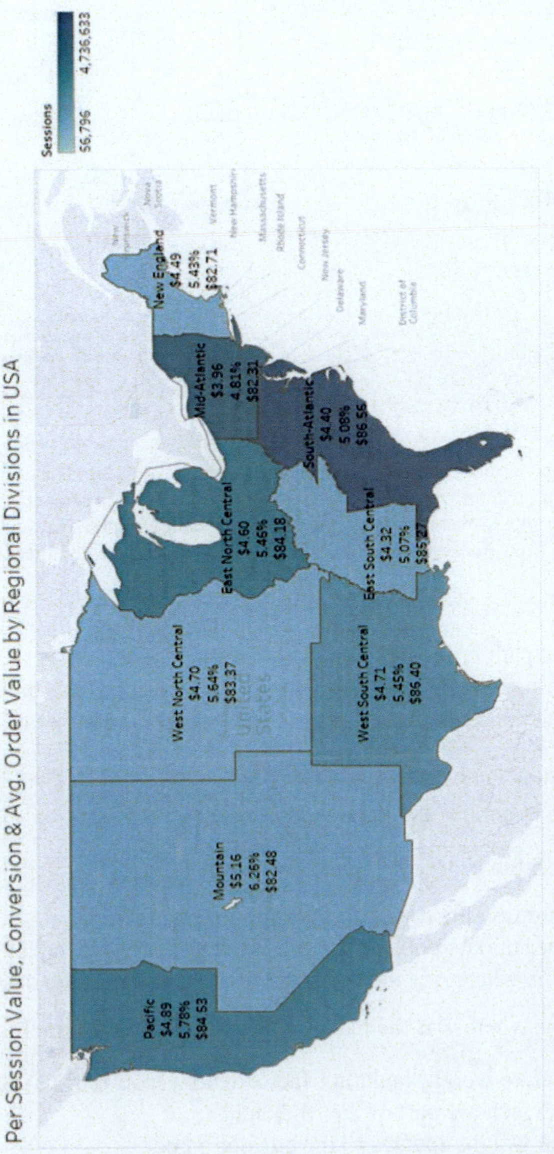

Fig. 3 Differences in per session value (i.e. revenue/num sessions), conversion, Avg. order value (i.e. revenue/transactions) across Regional Divisions in USA. Max differences observed are: Per Session Value– $5.16 versus $3.96, Conversion–6.26% versus 4.81%, Avg. order value–$86.56 versus $82.48

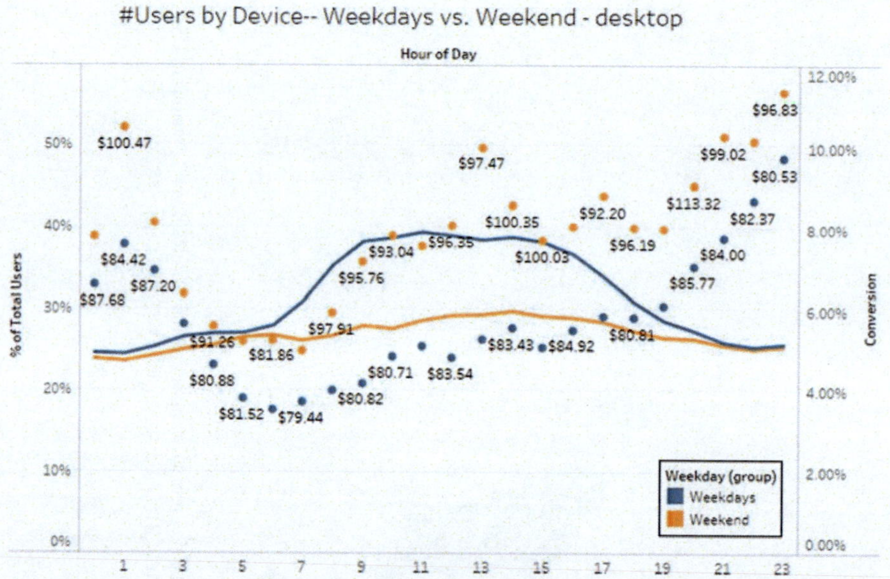

Fig. 4 Percentage of desktop users on weekdays versus weekend. The dots indicate conversion value (check *y*-axis to the right), whereas the text is the Avg. order value. What is evident is the decrease in desktop usage from morning to evening (work hours) on weekends compared to weekdays. Further, the jump in conversion from weekdays to weekends is accompanied by a significant jump in Avg. order value as well

- an existing model path or it can be empty
- a new set of training data on a periodical basis, where each data point represents an impression that consists of a set of attributes, the total number of impressions and the number of impressions with positive outcomes.
- model output path to save the new model
- configuration file to control the parameters.

The training algorithm proceeds as follows:

1. Read the configuration file to get all the training parameters.
2. Decay the number of data points/impressions in the existing model.
3. For $i = 1, 2, \ldots, n$, where n is the number of batches to split the data.

For $j = 1, 2, \ldots, m$, where m is the max number of iterations to run for each batch.

1. Calculate the feature weight updating factor with a map-reduce job using only the impressions which are part of the ith batch.
2. Apply the feature weight updates to the current model.
3. Check for model convergence and continue to next batch if converged.
4. Feature pruning based on the criterion mentioned above.

Some explanation about the algorithm:

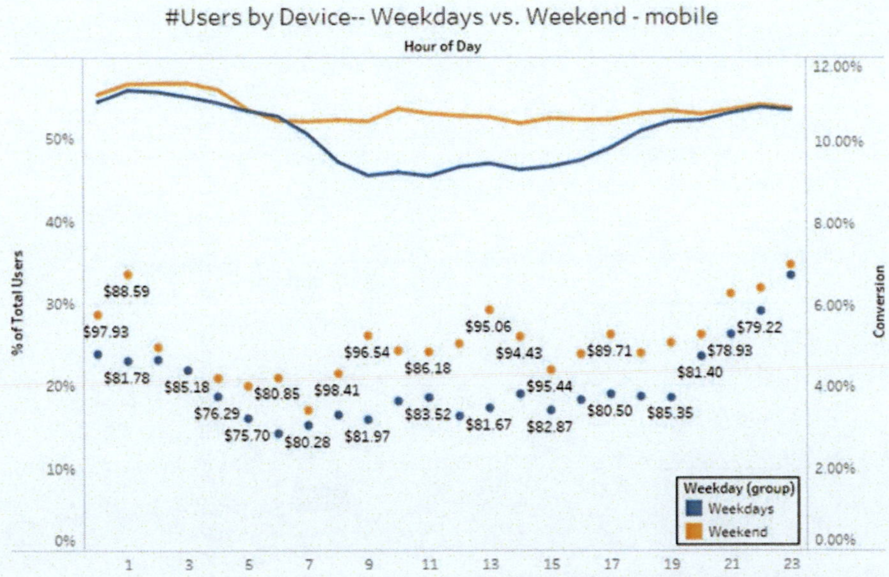

Fig. 5 Percentage of mobile users on weekdays versus weekend. The dots indicate conversion value (check *y*-axis to the right), whereas the text is the Avg. order value. What is evident is the increase in mobile usage from morning to evening (work hours) on weekends compared to weekdays. Further, the jump in conversion (or Avg. order value) from weekdays to weekends is not as strong as in desktop

- To determine a dynamic step size, we consider the number of impressions, while calculating the feature weight updating factor. Instead of using the raw counts, we want to weigh the recent events higher and therefore apply an exponentially decaying weights to the number of events.
- From Step 3, we know that the total number of map-reduce jobs is equal to the product of the number of batches and the number of iterations. If a batch converges before reaching the max number of iterations, we may have fewer jobs. If all the feature weight updating factors are close to 0, we consider the model to have converged with respect to the current batch of training data.
- The mapper class is designed to extract features from training data, make a prediction based on the current model, and output a weight updating factor for each feature.
- The reducer class is responsible for aggregating all the features stats. For each feature, the reducer outputs the sum of weight updating factors, the total number of events and the earliest timestamp.

The training parameters typically include the following:

- num_batches: number of batches we split the data into.
- max_num_iterations: maximum number of iterations we train for each batch.
- decay_factor: a time decay factor we apply to the existing model.

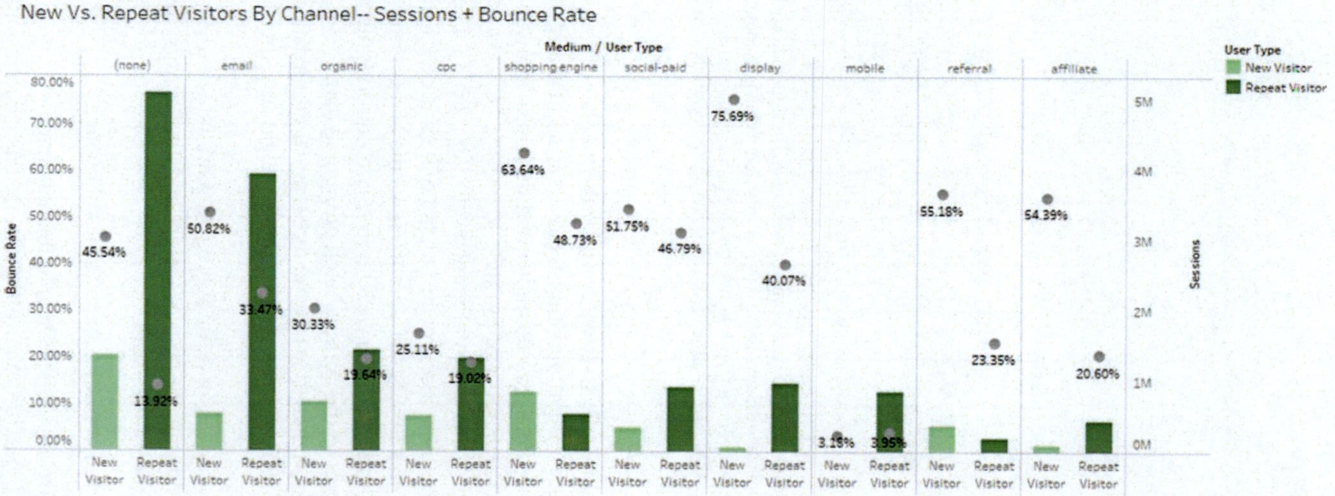

Fig. 6 Sessions and bounce rate by channel (filtered for top 10 across site) categorized by visitor type. Do note that channels referral (e.g., Facebook, YouTube) and shopping engine (e.g., Google, Bing) bring in more sessions from new visitors than repeat

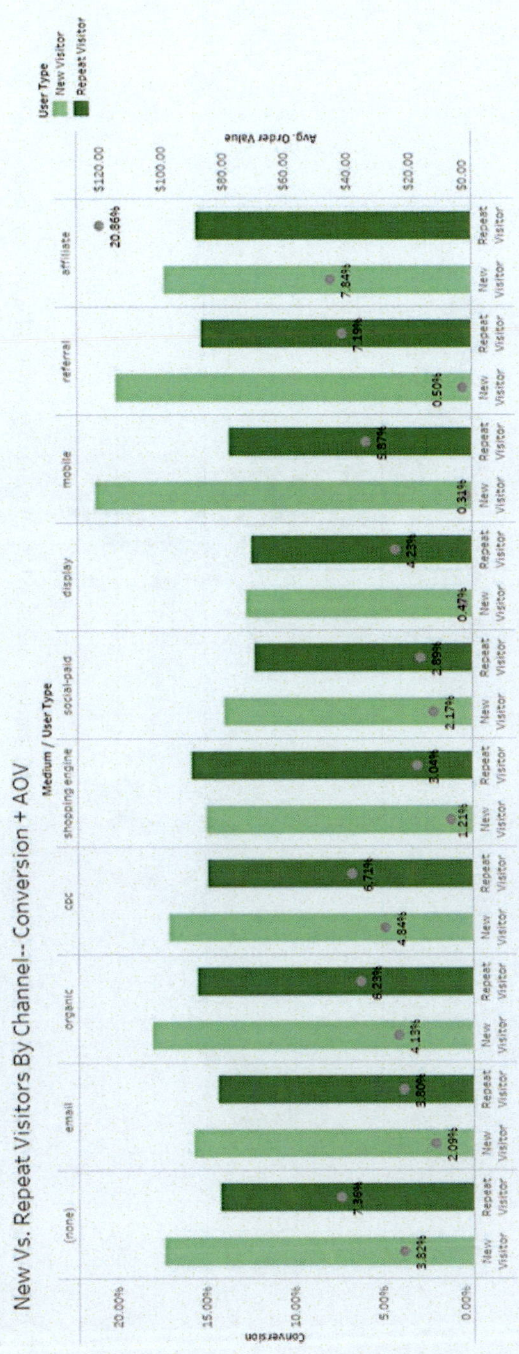

Fig. 7 Conversion and Avg. order value by channel (filtered for top 10 across site) categorized by visitor type. Do note that new visitors across all channels have higher Avg. order value

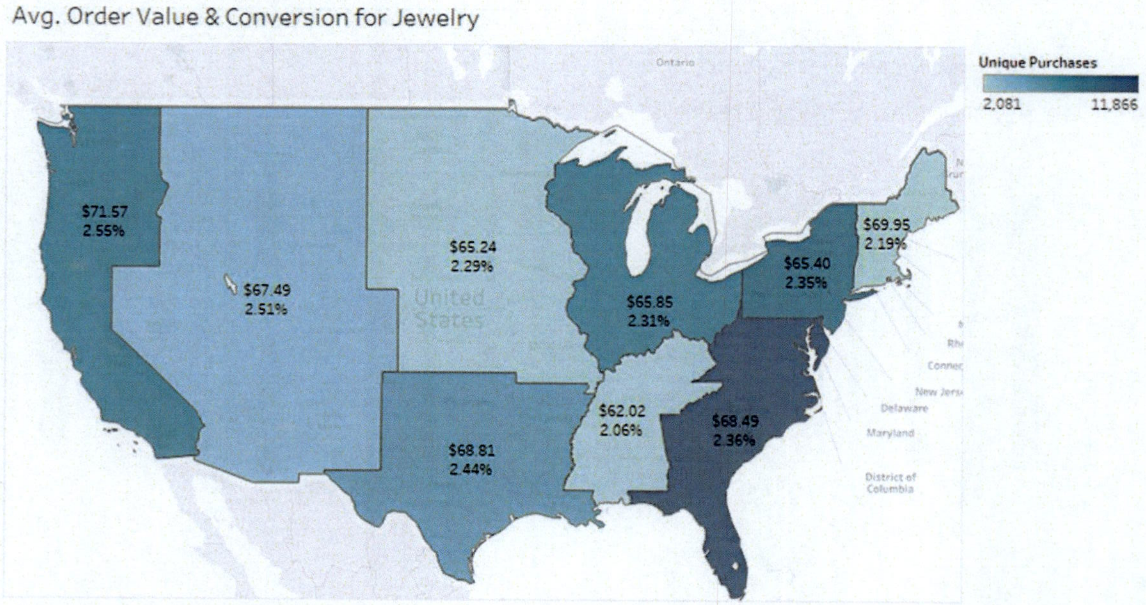

Fig. 8 AOV and conversion for jewellery (specifically necklaces, rings, earrings, beads and jewellery, etc.) products. Max differences observed: AOV– $71.57 versus $62.02, Conversion–2.55% versus 2.06%

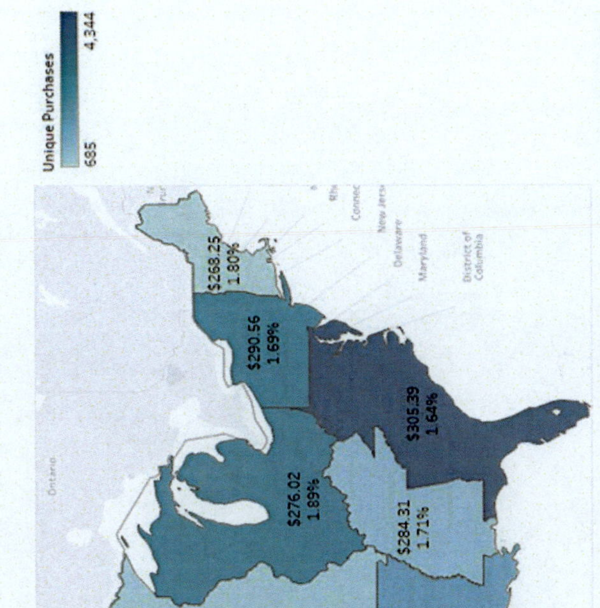

Fig. 9 AOV and conversion for electronics (specifically laptops, computers, computer accessories, TV accessories, cameras, etc.) products. Max differences observed: AOV–$327.93 versus $232.47, Conversion–2.02% versus 1.66%

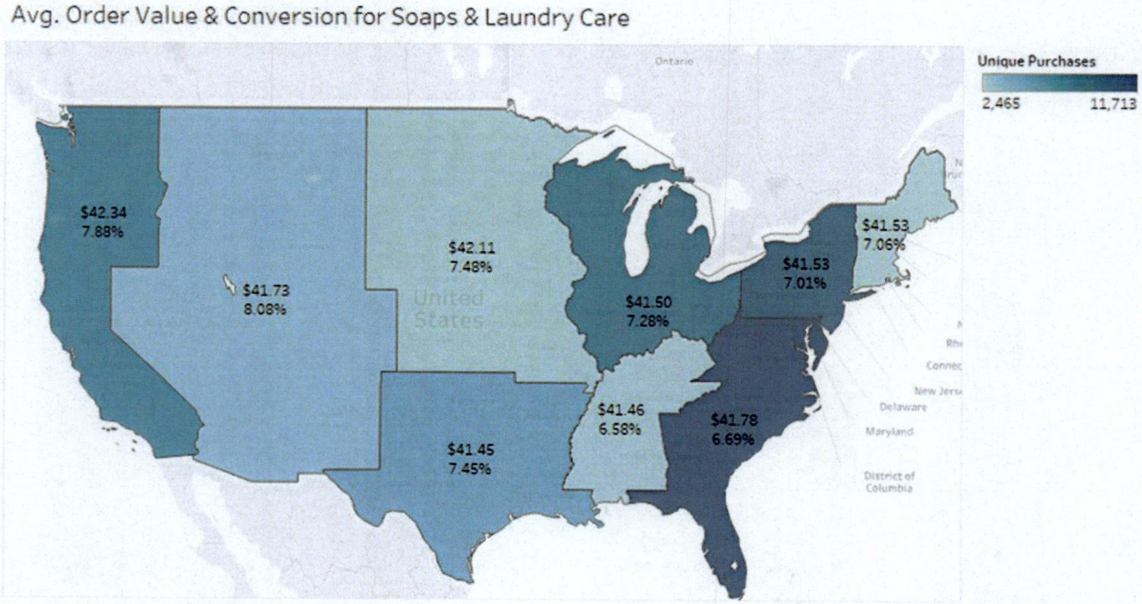

Fig. 10 AOV and conversion for soaps and laundry care products. Max differences observed: AOV–$71.57 versus $62.02, Conversion– 2.55% versus 2.06%

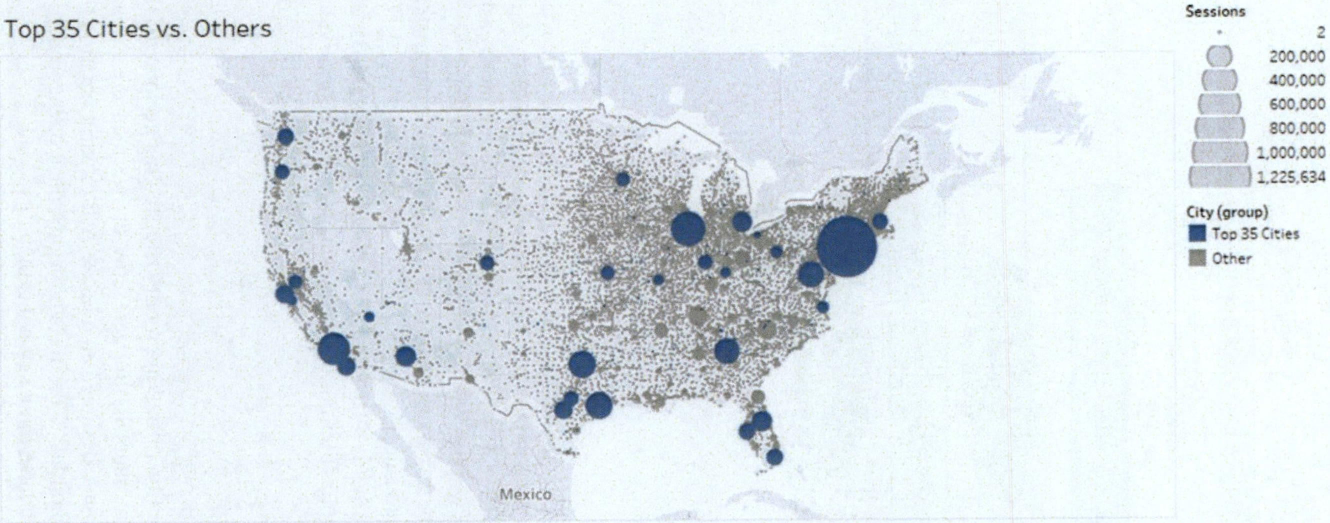

City (group)	Conversion	Per Session..	Avg. Order V..	% Mobile Se..	Avg. Session..	Bounce Rate
Top 35 Cities	3.93%	$3.50	$88.95	65.04%	5.81	33.44%
Other	5.79%	$4.83	$83.45	47.84%	6.94	28.18%

Fig. 11 Key performance indicators of top 35 cities versus rest

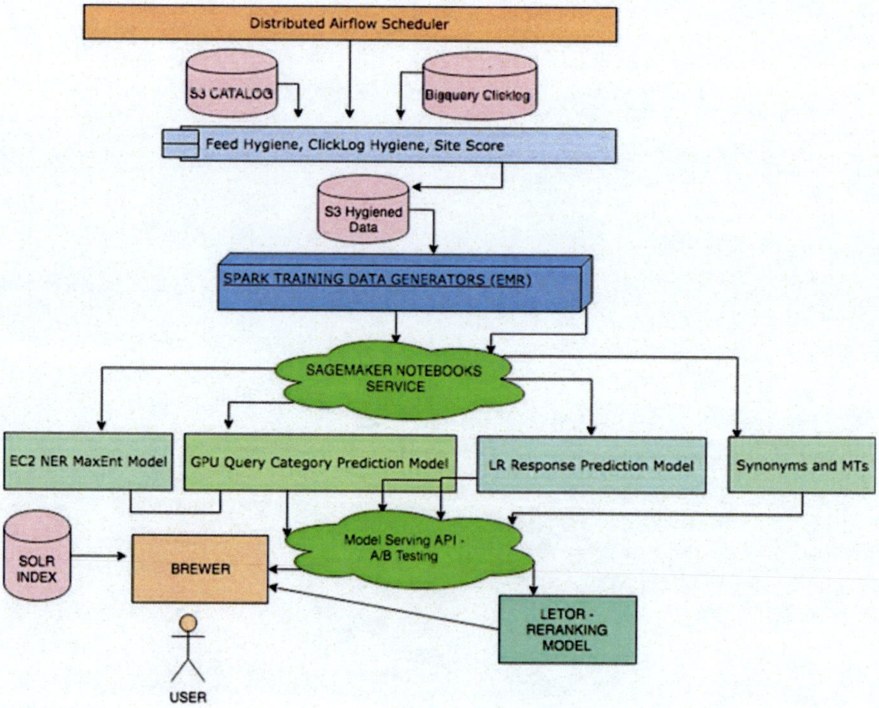

Fig. 12 ML architecture

- converge_threshold: a threshold to decide model training convergence.
- filtering_threshold: a threshold to filter out less frequent features.
- new_feature_bonus: allows new features to pass the filtering more easily.
- filter_by_total_events: whether to filter by total or average events.
- param_alpha, param_beta, param_gamma: learning step size parameters.
- prior: average response value.
- minimum_update_threshold: filter out updating factors below this threshold.
- minimum_update_events: filter out updating factors with too few events.

7 A/B Framework

When any response prediction or ranking model tries to change some of the system parameters, we need to measure its impact. How do we measure impacts of the model? To this effect, we have designed an inhouse A/B experimentation framework which works on these principles. Working within the space of incoming traffic and the system parameters, we have three key concepts:

- A domain is a segmentation of traffic.

- A layer corresponds to a subset of the system parameters or competing algorithms which have the same optimization criterion.
- An experiment is a segmentation of traffic where zero or more system parameters can be given alternate values that change the path how the incoming request is processed.

Domains contains layers and layers contains experiments. Domains let us have different partitioning of system parameters. For example, we can have a non-overlapping domain where we can change lots of parameters that might not normally be used together. Since a layer corresponds to subset of parameters that cannot be changed independently, experiments within a layer cannot run together for a request. Experiments from different layers however are free to run together on a request. Experiment allocation within a layer is based on some diversion type available in the request. This is to ensure stickiness with respect to that type. Diversion type can be device ID or browser cookie or query string. Experiments that target user behaviour would usually want that the users do not pop in and out of the experiment and therefore need user stickiness. Some experiments do not need user level stickiness but would operate at query level and therefore would want stickiness at query level. We partition the traffic space into say 1000 buckets and assign experiments within a layer to bucket ranges. The request is mapped to a bucket using a function of the bucket ID and layer modulus the number of buckets, e.g., f(id, layer) % 1000. Note that this function takes layer as an argument to ensure that the experiments in different layers are independently diverted.

The graph below, Fig. 13 shows the result of A/B experimentation between a control group and a response prediction model group. Unbxd clearly shows an uptick of 10% in CTR for torso and tail queries in terms of search volumes (where torso and tail refers to terciles of search volume distribution) and AOV for an American site where the users where shown search results ranked by the response predictor algorithm by the probability of click.

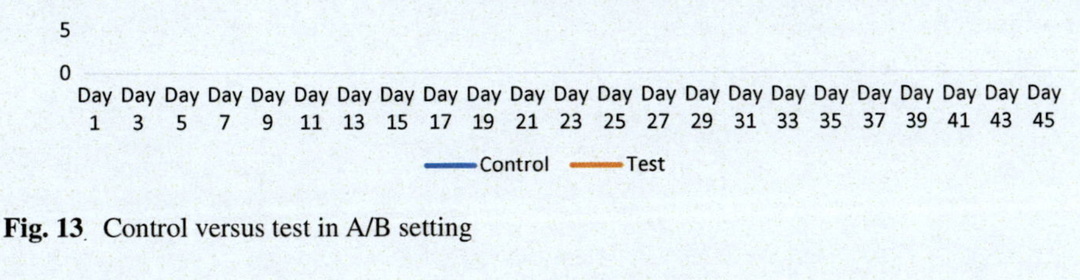

Fig. 13 Control versus test in A/B setting

8 Future Work

From the above feature-based response prediction model, we have been able to both personalize the search results for a user given a query and improve the CTR and CVR which is the revenue tracking metrics of the search business. Through feature-based ranking models, we have mostly captured the general trends of clickability of a product and improved the business performance metrics by driving a higher average order value. However, we have not explored the option of serendipity or cross learning when it comes to surprising the user or providing related product recommendation in the same search session. Window shopping and serendipity shopping is another paradigm which is also known to improve engagement of a shopper and a site. In literature, cross-selling products, "bought also bought", "viewed also viewed" are common basis for recommendations. In search however since user's context is set through a query, we cannot drift afar, but using an epsilon greedy approach or multi-armed bandits, we can exploit our feature-based predictions and explore with a subtle mix of random predictions. This would be the next set of ranking algorithms that we look forward to working on in the future.

References

Cheng, H., & Cantu-Paz, E. (2010). Personalized click prediction in sponsored search. https://www.wsdm-conference.org/2010/proceedings/docs/p351.pdf.

Guo, H., Tang, R., Ye, Y., Li, Z., & He, X. (2017). DeepFM: A factorization-machine based neural network for CTR prediction. https://arxiv.org/pdf/1703.04247.pdf.

Kumar, R., Kumar, M., Shah, N., & Faloutsos, C. (2018). Did we get it right? Predicting query performance in e-commerce search. https://sigir-ecom.github.io/ecom18Papers/paper23.pdf.

Tze Cheng, H., Koc, L., Harmsen, J., Shaked, T., Chandra, T., Aradhya, H., et al. (2016). Wide and deep learning for recommender systems. https://arxiv.org/pdf/1606.07792.pdf.

Zhou, G., Song, C., Fan, Y., Zhu, X., Zhu, H., Ma, X., et al. (2018). Deep interest network for click-through rate prediction. https://arxiv.org/pdf/1706.06978.pdf.

Connectedness of Markets with Heterogeneous Agents and the Information Cascades

Avijit Ghosh, Aditya Chourasiya, Lakshay Bansal, and Abhijeet Chandra

1 Introduction

Macroeconomic integration of global financial markets is often characterized as complex systems where ever-increasing interactions among a vast number of agents make it difficult for the traditional economic theory to provide a realistic approximation of market dynamics. Schweitzer et al. (Sci., 325: 2009) emphasize that economic systems are increasingly interdependent through cross-country networks of credit and investment, trade relations, or supply chains, and highlight the need for an integration of network theory and economic models to reduce the risk of global failure of financial systems.

The diversification argument suggests that price fluctuations attributed to macroeconomic shocks tend to average out over time and, therefore, have little aggregate effects on the system. However, this argument ignores the fact that economic agents operating in financial markets are interconnected. This connectedness might propagate idiosyncratic shocks throughout the financial system. Such "ripple effects" or "cascade effects" are observed during the global financial crisis of 2007–08 (Mulally 2008).

A. Ghosh (✉) · A. Chourasiya · L. Bansal · A. Chandra
Vinod Gupta School of Management, Indian Institute of Technology Kharagpur, Kharagpur, India
e-mail: avijitg22@gmail.com

A. Chourasiya
e-mail: adityachourasiya142@gmail.com

L. Bansal
e-mail: lakshayb5@gmail.com

A. Chandra
e-mail: abhijeet@vgsom.iitkgp.ac.in

A. K. Laha (ed.), *Applied Advanced Analytics*, Springer Proceedings in Business and Economics, https://doi.org/10.1007/978-981-33-6656-5_18

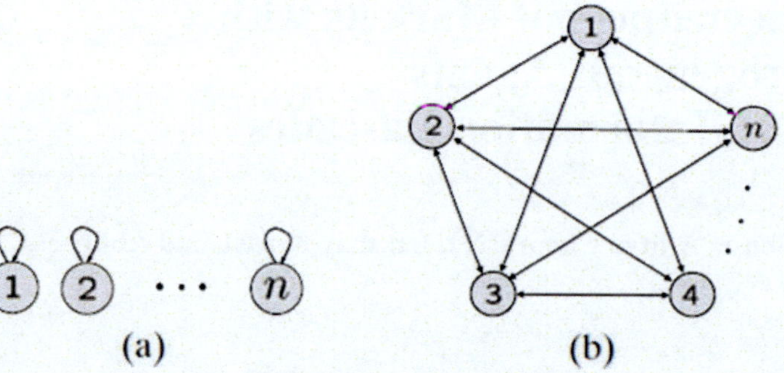

Fig. 1 Sample representations of the network of two financial systems that are symmetric. **a** A financial system with no economic agent depending on other agents. **b** A financial system with each agent relying equally on all other economic agents (Adopted from: Acemoglu et al. Econometrica 2012)

Connectedness among firms in a financial market becomes more prominent issue with respect to risk management and its key aspects such as market-level risk in terms of return connectedness and portfolio concentration, credit risk in terms of default connectedness, and systemic risk in terms of interconnectedness among firms (Diebold and Yilmaz 2014). It is, therefore, a central issue to understand how different agents in a financial market are connected to each other and how they share risks. The dependence among the agents could be either flat-structured where each agent acts independently and in the process contributes to the overall system in unique ways (see Fig. 1a), or interconnected as a web where each agent contributes not only to the overall system but also to the shocks caused by other agents (Fig. 1b). The connectedness of the economic agents is therefore important to study as it helps to identify potential weak agents and understand the source of cascading defaults and failures well in advance and thereby ensure that the system operates well.

In this paper, we develop a model for the diffusion of shocks across heterogeneous yet interconnected macroeconomic agents in terms of a basic growth-based network. Specifically, our model provides new evidence regarding the co-integration of select major world economies and sharing of risks in terms of cascades of failures across the network of the economies. The interdependence of firms and institutions in a network can be explained with the help of cross-holdings of assets and liabilities. Suppose a firm loses its economic value for certain reasons, intrinsic or extrinsic, and touches a downside threshold that further propagates a fall in economic value of the firm, it will have economic implications for other firms operating in the network. Since the firms operating within a network directly or indirectly cross-hold assets and liabilities, the losses in economic value of one (or more) firm(s) will propagate loss in economic value of other firms as well. Even though some firms might not transact directly with the firm(s) that loses economic value, it will be affected as a result of being part of the broader network. By this way, risks of firms in a network are shared and propagated by other firms operating in the network. Interestingly, at every stage, some firm may touch a threshold and lose value discontinuously. This may, at macro-level, even amplify the risks in the network substantially.

In our work, we consider forty-three major world countries covering about 85% of world gross domestic product (GDP) as heterogeneous economic agents that hold economic assets (e.g., any factor of production or other investment) as well as liabilities. The primitive holdings of these agents are represented in terms of input–output of these countries obtained from World Input-Output Database (WIOD: Trimmer et al., Rev. Int. Eco. 2015). Our network model shows that cascades of failures spread rapidly in the network and that large markets with values above certain threshold are able to sustain the cascades of failure much longer compared to the smaller markets even though these smaller markets have value slightly below the threshold. The information about the threshold of economic value, below which the cascade of failure starts in the network, for different countries can help us avoid failure of the network as a whole. We also show that the connectedness of markets is well represented in the WIOD datasets over time. Our backbone network model narrows down the connectedness of the markets to show the dynamic impact of industry-specific input–output on the network formation in our sample.

The remaining of the paper is organized as follows. Section 2 discusses the relevant and recent literature on the applications of networks in finance and macroeconomic research. Section 3 presents the details pertaining to the data, their characteristics, sources, and mathematical and empirical methods to implement the network approach to macro-finance issues in the context. Preliminary results obtained through the analyses of the data are provided in Sect. 4, followed by detailed discussions and the implications of the results in Sect. 5. The paper ends with a summary and concluding remarks along with the issues that are left unaddressed in this work in Sect. 6.

2 Related Work

Financial markets in general and stock markets in particular act as complex systems consisting of several agents that interact with each other in a stochastic manner. The level of complexity is attributed to many factors including the structure of micro- and macro-environments, the heterogeneity of the participating agents, and their interactions with each other.

Economic agents such as firms at micro-level and countries and/or markets at macro-level are interconnected by way of international trade relations, eco-political integration, and information spillovers. They share risks and shock as well as get shocked by each entity in the system. Recent literature dealing with issues such as contagion and cascades of failures across multiple agents in a system highlights the context in which agents are ex ante heterogeneous. In such cases, the risk characteristics of the shocks in the system hit the projects of the various agents. Because of the increase in interconnections among firms and/or markets (Diebold and Yilmaz 2014), we argue that it is inadequate to consider the regulation of capital requirements and macro-prudential investigation as reliable. Some researchers have suggested that a policy that allows modulation exposures within a network can act as an effective tool (Stiglitz 2010) that are evident in a concrete empirical framework as well (Degryse and Nguyen 2007).

One customary way of addressing the problem of macro-prudential regulation is to rely on stress tests. Local macro-prudential policy in a core country tends to affect the cross-border transmission of local macroeconomic policies by way of lending abroad, by restricting the increase in lending by less strongly capitalized banks in the country. This essentially requires the researchers and policy makers to carry out empirical studies that examine the outcomes for a system in which institutions that are interconnected are subjected to large shocks (see, e.g., ESMA (2006); Kara et al. (2015)). This, however, does not consider that a well-connected network will likely collapse after subjected to a sufficiently large shock. It is argued that such connectedness at times offers the benefit of forestalling problems when the network experiences smaller yet frequent shocks. It is, therefore, not feasible to achieve a certain level of the optimality of a particular connection structure unless we incorporate the impact of the whole distribution of shocks in the system. An alternative approach for assessing the risks in the banking systems is to hypothetically simulate the impact of stock drawn from the empirical distribution of the historical returns under the connectedness as seen in the actual network (Elsinger et al. 2006).

Moreover, examining the information spillovers across markets/firms can highlight similar insights. Preliminary research within the domain of risk spillover and market contagion in the debt markets emphasizes on an examination of primary determinants of debt markets. These studies use structural, financial, institutional indicators such as micro-factors, and macroeconomic characteristics that explain the dynamic movements of the yields of a sovereign bond. In this context, seminal works by Eichengreen and Luengnaruemitchai (2004), Claeys and Vašíˇcek (2014), and Burger and Warnock (2006) for the Asian and European markets; Eichengreen et al. (2004) empirical work in the Latin American markets; Adelegan and Radzewicz-Bak (2009) and Mu et al. (2013) for the African markets emphasize on the significance of macro-economic factors in determining the spillovers across markets. These studies argue that the volatilities in the exchange rate and the fiscal characteristics typically hinder the development of both sovereign and corporate bond markets across most of the emerging economies. On the contrary, institutional, firm-specific, and structural characteristics, such as trade openness and bureaucratic qualities, provide a positive nudge for the growth of both sovereign and corporate bond markets. In this context, the application of network approach to understand the contagion and risk sharing attributes of different markets is studied by Ahmed et al. (2018).

The literature on financial contagion, cascades of failures, and systemic risk spans across markets and economies and appears to be emerging steadily in the recent past. It captures the imagination of researchers across disciplines including those primarily working in financial economics, mathematics and other computational domains, and engineering disciplines. Hence, we present only a brief summary of some of the more closely related and recent research works.

The research on interconnectedness was pioneered by Allen and Gale (2000) who studied the stability in interconnected financial systems in a developed market.

They propose a model in line with Diamond and Dybvig (1983), where a network structure has a single and completely connected component. This structure is always optimal. Such a network minimizes the extent of default. We show in our model the contrasting results. We find that a richer shock structure generates a genuine trade-off between the risk shared within the network and the contagion effect and that both segmentation and lower dispersion of connections may be optimal for the network.

Some more recent, related research on the issue is by Elliott et al. (2014), Diebold and Yilmaz (2014), Glasserman and Young (2015), Acemoglu et al. (2015), and Ahmed et al. (2018). The nature and form of the financial linkages among firms (and in some contexts, markets) examined in these works are seemingly different, as they consider both the asset side and the liability side of the firms' (country's) balance sheet. Such a framework, in turn, entails the presence of a mechanism that amplifies the shocks hitting a firm (a country) and subsequently spreading across the network in unique manners.

In a multi-firm context, literature suggests the presence of the trade-off between the risk distribution enabled by stronger interconnection and the increased exposure to cascades as an outcome of larger components in the financial network (Cabrales et al. 2013). Cabrales et al. (2013) study selected benchmark networks that are minimally interconnected and complete, to identify the best for different distributions of shocks.

Another unique approach that our work focuses on is related to the implementation of network formation in growth model framework and the related analysis. Elliott et al. (2014) characterize conditions for the macroeconomic structure of the network under which the default cascades might occur. We, however, aim to characterize the optimal and dynamic structures of financial and economic networks in diverse scenarios and also investigate if the industry context matter for the network sustains the shocks. In an earlier examination, Shaffer (1994) too suggests a moderated relationship between risk spillovers and systemic failures of economies. Although entities hold diversified portfolios to reduce risk, they also face the risk of owning similar portfolios in the market and being in a system that might be susceptible to contemporaneous failures. Acemoglu et al. (2015) highlight the optimal structure of financial networks, but they focus on examining the shock distributions that are concentrated within a system for a given shock magnitude. More recent works, on the contrary, present the properties of the curvature of the function of the risk exposure and the cumulative distribution of shocks (Cabrales et al. 2017). These more recent evidences with regard to the network formation among heterogeneous economic agents allow us to incorporate in our analysis a rich set of possible shock distributions. We can then show different ways of variations in an optimal financial structure, in response to the characteristics of those shock distributions. This also enables us to examine the dynamic properties of the networks over time and other inputs such as industry category. This uniqueness of our work is a significant contribution to the theoretical and empirical work on the issue.

Another line of the literature highlights how financial contagion and cascades of failures are affected by imperfect information about the shocks hitting the system. This is studied widely in econometrics and financial economics literature on information spillover and market co-integration to certain extent. Some recent work, for example, Allen et al. (2012), studies the effects of the arrival of a signal on segmented and unsegmented structures of the network. These signals indicate that a firm in the system will have to default. This argument can further be extended to the networks of markets from different countries.

Finally, it is important to discuss the empirical and policy-oriented evidence that has been the main objective of bringing in the summary measures for the network connectedness. These measures are derived from the network of relationships among business entities (mostly financial firms) with the aim of predicting the probabilities of systemic failures. Some studies, for example, propose different measures of centrality in networks (Battiston et al. 2012; Denbee et al. 2011). A significant contribution in this respect is the work by Elsinger et al. (2006), who use data from the Austrian market and show that a correlation in banks' asset portfolios can be considered as the main source of systemic risk.

3 Research Objectives

The above-mentioned review of relevant research examining twin issues of the interconnectedness of heterogeneous economic agents and the contagion and cascades of failures across firms and markets, information and risk spillover, and macro-prudential regulations in the context of financial and economic entities such as firms and markets has brought out the following research issues to be examined:

- Since the theories emphasize that the economic entities such as firms and/or markets, whether homogeneous or heterogeneous, are interconnected to certain extent, how do these entities form networks? It would be of empirical interest to investigate how these entities connect to each other in terms of the directional spillover of risks and cascades of failures.
- With a growth model framework, the network formation tends to evolve with change in underlying attribute(s), such as time, values, and so on. We propose to examine how industry-specific inputs affect the formation of networks of economic entities. This would suggest the interconnectedness of economic entities at the micro-level where firm/market characteristic(s) becomes an important input to identify the potential risk in the network.

4 Methodology

Our aim is to study the cross-holdings of entities in terms of input–output and look at a time-varying feature to examine the changes in the network. We also hope to study the ripple effects caused due to the failure of entities inside the model. It is hypothesized that the ripple effects in the network should be caused once an identified entity(-ies) touches or crosses the threshold indicating the cascade of failures.

4.1 Data Source and Details

To measure the value of cross-holdings of different economic entities, our study focuses on tracking the flows of products across industries and countries. Our main measure is the World Input-Output Database (WIOD). This data, available at the database Web site (http://www.wiod.org), which provides for the data on input–output for several countries, has been specifically constructed for studying the relationship among countries through trade flows (see Timmer et al. 2014, 2015; Dietzenbacher et al. 2013). The database also provides world input–output data for each year starting from 1995 of more than forty countries. The countries included in the database are all twenty-seven countries of the European Union (as of January 1, 2007) and 15 other major countries, namely China, Brazil, India, Australia, Canada, Mexico, Japan, Indonesia, Russia, Taiwan, Turkey, South Korea, and the USA. These 43 countries are economically significant in the world trade ecosystem as they represent more than 85% of world GDP. In addition, we also incorporate a model for the remaining non-covered part of the world economy. This model is designed such that the decomposition of final output is complete for value addition. This model captures thirty-five industries spanning across the overall economy, including sectors such as agriculture, utilities, construction, and mining, fourteen manufacturing industries, and seventeen services industries. Our sample dataset comprises the WIOD data for all years between 2000 and 2014 for all the 43 countries and industries available in the database.

4.2 Estimation Framework

Our framework proposes that there are n organizations (as economic agents such as countries, financial firms, or other business entities) making up a set $N = 1, ..., n$, with $n = 43$. The values assigned to these sample organizations are ultimately based on the economic value of asset holdings or factors of production—henceforth, simply assets $M = 1, \dots, m$. For consistency, an asset holding may be taken as a project that is expected to generate a series of cash flows over time. The present value (or the current market price) of asset k is denoted p_k. Further, let $D_{ik} \geq 0$ be the share of the value of asset k owned by an organization i that receives the cash flows and let $\mathbf{D}$ denote the matrix whose (i, k)th entry is equal to D_{ik}.

An organization can also hold shares (here we have an amount of debt held by one country from another country) of other organizations in the sample. For any $i, j \in N$, the number $C_{ij} \geq 0$ is the fraction of the organization j owned by the organization i, where $C_{ii} = 0$ for each i. The matrix $\mathbf{C}$ can be proposed to be a network with a directed link from i to j, if i holds a share of j with a positive value, so that $C_{ij} > 0$.

After we account for all these cross-holding shares across sample organizations, we are left with a share $\widetilde{C_{ii}} := 1 - \sum j \in N_i C_{ij}$ of organization i that is not owned by any organization in the system. This component of the share is assumed to be of positive value. Theoretically, this is the part that is held by outside shareholders of the organization i, and is external to the system of cross-holdings. The off-diagonal entries of the matrix $\widetilde{\mathbf{C}}$ are defined to be 0.

The equity or book value V_i of an organization i is the total value of its shares. The value is obtained by adding the value of the shares owned by other organizations and the shares owned by outside shareholders. This value equals to the value of organization i's asset holding plus the value of its claims on other organizations in the system:

$$V_i = \sum_k D_{ik} p_k + \sum_j V_j C_{ij} \tag{1}$$

In matrix notation, the above equation can be written as

$$\mathbf{V} = \mathbf{Dp} + \mathbf{CV} \quad or \quad \mathbf{V} = (\mathbf{I} - \mathbf{C})^{-1}\mathbf{Dp} \tag{2}$$

As shown in both Brioschi et al. (1989) and Fedenia et al. (1994), the market value reflects the external asset holdings. The final non-inflated economic value of an organization to the economy is well captured by the equity value of that organization that is held by its outside investors. This economic value captures the flow of real assets that is expected to accrue to the ultimate investors of that organization. The market value is denoted by v_i and equals to $C_{ii} V_i$, and therefore:

$$v = \widetilde{\mathbf{C}}V = \widetilde{\mathbf{C}}(\mathbf{I} - \mathbf{C})^{-1}\mathbf{Dp} = \mathbf{ADp} \tag{3}$$

where

$$\mathbf{A} = \widetilde{\mathbf{C}}(\mathbf{I} - \mathbf{C})^{-1} \tag{4}$$

Here, $\mathbf{A}$ is the *dependency* matrix. Suppose in our system, every organization holds the ownership of exactly one unique asset, so that $m = n$ and $\mathbf{D} = \mathbf{I}$. We further propose that A_{ij} describes the dependence of the value of the organization i upon the organization j's unique asset holding. It is then logical to propose that $\mathbf{A}$ is column-stochastic, so that the total economic values of all organizations supposedly add up to the total economic values of all underlying assets—then for all $j \in N$, we have

$$\sum_{j \in N} A_{ij} = 1 \quad (5)$$

If the market value v_i of an organization i in the system, for any reason, drops below a threshold μ, then i is said to fail, in economic sense, and incurs failure costs $\beta_i(p)$. These failure costs are then subtracted from the cash flows received by the failing organization. In such scenario, these organizations can propose to push the diversion of cash flow to deal with the failure. They can also hope a decrease in the returns that the organization generates from the unique assets that they hold. Either way, the proposed approach introduces critical nonlinearities, or rather discontinuities, into the system of organizations.

We have calculated a *fractional ownership* matrix which represents the fraction of GDP produced by the country using its own resources and debt taken from other countries. Here, GDP is the total output from all industries, where each industry might or might not have borrowed money from other countries. We have taken the base year as 2000. We have normalized the GDP values with the GDP of India.

So, if we consider all the diagonal elements of a *fractional matrix* as zero we get a **C** matrix and by forming a matrix with the diagonal elements of the *fractional matrix* give a $\widetilde{\mathbf{C}}$ matrix (non-diagonal elements are filled with zeros). Using Eq.(4), we can calculate matrix **A**.

We define parameter $\theta \in [0, 1]$, which is used to calculate the fractional decrease in GDP values of base year which can cause the country to fail, and numerically threshold is defined as:

$$\Upsilon = \theta * (\mathbf{A}.\mathbf{p}_t) \quad (6)$$

Here, p_t is the normalized GDP of base year (2000). So, if the normalized GDP of any country goes below this threshold we consider that country to be a failure. If any country fails, we subtract 50% of the threshold value from its normalized GDP of the present year and repeat the process again until no country fails.

Mathematically, steps can be visualized as:

1. $\mathbf{A} = \widetilde{\mathbf{C}}(\mathbf{I} - \mathbf{C})^{-1}$ (nXn matrix)
2. $\mathbf{p} = GDP$ (Normalised nX1 matrix)
3. $\mathbf{p}_t = GDP$ (Normalised base year (2000) nX1 matrix)
4. $\Upsilon = \theta * (\mathbf{A}.\mathbf{p}_t)$
5. If $\mathbf{A}.\mathbf{p} < \Upsilon$ (compared elementwise) follow step 6 else all countries are safe at that Θ.
6. $p = p - \Upsilon/2$ (elementwise, subtract $\Upsilon/2$ only from countries which has failed)
7. Repeat step 5 until no country fails.

To get a qualitative idea of the accuracy of our network, we run the cascading model on the WIOD dataset from 2000 to 2014, and try to see if there are economic explanations for failure of the countries reported by the algorithm. Once the veracity of the algorithm is ascertained, we will look into simulating a large amount of trading and then repeatedly apply our failure model to find the most ideal trading scenarios for the least failures.

5 Preliminary Results

The WIOD dataset provides the input–output in current prices and is denoted in millions of dollars. The WIOD database covers twenty-eight countries from the European Union and fifteen other major countries (totaling to 43 countries) from across the world for the sample period from 2000 to 2014. The data is sorted industry-wise for 56 industries in total. For a preliminary analysis, we have condensed the data to only the inter-country input–output table. The interlink, with respect to trade, between the countries is given in Fig. 2.

6 Discussion

Traditional approach to understand the way in which an economic agent's—an individual or a firm in particular and a market or an economy in general—behavior depends on what other economic agents do involves insights from cognitive and mathematical sciences, such as psychology and game theory among others. Researchers

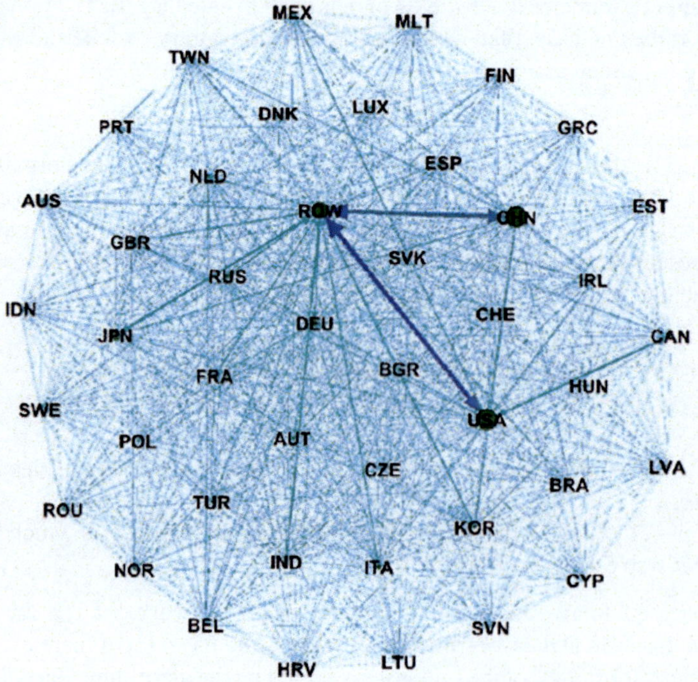

Fig. 2 A representative network diagram of the sample countries based on the WIOD dataset for the year 2016

carry out controlled and non-controlled experiments to observe and analyze the reaction of economic agents as a response to shifts in behavior of other players. This mechanism explains the nature of interconnectedness of agents in a system where the agents are believed to be part of the network and hypothesized to be homogeneous. Recent research focuses on heterogeneity characteristic of these economic agents to understand the nature of interconnectedness in the network.

In finance, the interconnectedness becomes more significant as it helps understand the contagion of risk, the flow of information (that eventually explains the price discovery process and existence of arbitrage opportunities), and overall the ability of the system to absorb the shocks caused by one or more agents (Diebold and Yilmaz 2014). During the 2008 global financial crisis, several firms belonging to the finance industry across the globe were so deeply interconnected that one failure in one part of the world would result in tremors across several players in the industry. For example, the fall of Lehman Brothers caused few big finance firms to become susceptible to bankruptcy that led the government to swing into action and bail out many other financial institutions to save the entire economy (or, probably several economies around the world). In this context, examining the interconnectedness of the markets should explain the vulnerability of the system and pinpoint the weak nodes in networks so that regulators and governments among others take corrective actions well in time.

7 Concluding Remarks

Our research presents the evidence on the nature of interconnectedness that global markets exhibit in terms of their input–output representing the cross-holdings. It shows that the interdependence of some markets in a global network has strong correlation with not only the size of the markets, but also the direction of trades/cross-holdings, and the type of industries that dominate in their input–output data. With growth model estimation, we are able to project the cascades of failures in the network significantly. Our results as exhibited in the graphs corroborate with the empirical research on the failures of the markets. It is shown that markets having more connections with other markets in the network are likely to sustain the shocks and cascades of failures for longer time. This evidence is aligned with the argument of diversification as a strategy to mitigate risk. Our findings employ innovative approaches such as network formation approach and graph theory to explain the interconnectedness of markets across the world, and contribute significantly to the theoretical issues related to market integration and risk spillover (Diebold and Yilmaz 2014; Cabrales et al. 2017; Ahmed et al. 2018) (Tables 1, 2, 3, 4, 5 and 6).

Acknowledgements The authors are grateful to the reviewers and the editor for their comments and suggestions. The financial support of Indian Institute of Technology Kharagpur through the Challenge Grant (CFH ICG 2017 SGSIS/2018-19/090) toward this research is hereby acknowledged. Usual disclaimers apply.

8 Appendix

Table 1 Name of countries failed at that threshold and the cascade impact in 2001

	2001
0.1	[[]]
0.775	[[]]
0.78	[['TUR'] []]
0.785	[['TUR'] []]
0.88	[['TUR'] []]
0.885	[['BRA' 'TUR'] ['TWN'] ['JPN'] []]
0.89	[['BRA' 'JPN' 'TUR' 'TWN'] []]
0.91	[['BRA' 'JPN' 'TUR' 'TWN'] []]
0.915	[['BRA' 'JPN' 'TUR' 'TWN'] ['MLT'] []]
0.92	[['BRA' 'JPN' 'TUR' 'TWN'] ['KOR' 'MLT'] []]
0.925	[['BRA' 'JPN' 'TUR' 'TWN'] ['KOR' 'MLT'] []]
0.93	[['BRA' 'JPN' 'TUR' 'TWN'] ['KOR' 'MLT'] []]
0.935	[['BRA' 'JPN' 'KOR' 'TUR' 'TWN'] ['AUS' 'MLT'] ['IDN'] []]
0.94	[['BRA' 'JPN' 'KOR' 'MLT' 'TUR' 'TWN'] ['AUS' 'IDN' 'SWE'] []]
0.945	[['BRA' 'JPN' 'KOR' 'MLT' 'SWE' 'TUR' 'TWN'] ['AUS' 'IDN'] []]
0.95	[['BRA' 'JPN' 'KOR' 'MLT' 'SWE' 'TUR' 'TWN'] ['AUS' 'IDN'] []]
0.955	[['BRA' 'JPN' 'KOR' 'MLT' 'SWE' 'TUR' 'TWN'] ['AUS' 'IDN'] ['ROW'] ['FIN' 'LUX'] []]
0.96	[['AUS' 'BRA' 'JPN' 'KOR' 'MLT' 'SWE' 'TUR' 'TWN'] ['FIN' 'IDN' 'ROW'] ['CYP' 'LUX'] []]
0.965	[['AUS' 'BRA' 'JPN' 'KOR' 'MLT' 'SWE' 'TUR' 'TWN'] ['FIN' 'IDN' 'ROW'] ['CYP' 'LUX'] []]

Table 2 Name of countries failed at that threshold and the cascade impact in 2002

	2002
0.1	[[]]
0.95	[[]]
0.955	[['JPN'] ['BRA'] []]
0.96	[['BRA' 'JPN'] []]
0.975	[['BRA' 'JPN'] []]
0.98	[['BRA' 'JPN'] ['ROW'] []]
0.985	[['BRA' 'JPN'] ['ROW'] ['TWN'] ['USA'] ['CAN' 'MEX'] []]
0.99	[['BRA' 'JPN'] ['ROW'] ['TWN' 'USA'] ['CAN' 'MEX'] []]
0.995	[['BRA' 'JPN' 'ROW'] ['CAN' 'TWN' 'USA'] ['MEX'] []]

Table 3 Name of countries failed at that threshold and the cascade impact in 2009

	2009
0.1	[[]]
0.725	[[]]
0.73	[[]]
0.735	[['LTU'] []]
0.74	[['LTU'] []]
0.745	[['LTU'] []]
0.75	[['LTU' 'LVA'] []]
0.755	[['LTU' 'LVA'] []]
0.76	[['LTU' 'LVA'] ['EST'] []]
0.765	[['EST' 'LTU' 'LVA'] []]
0.77	[['EST' 'LTU' 'LVA'] []]
0.775	[['EST' 'LTU' 'LVA'] []]
0.78	[['EST' 'LTU' 'LVA'] []]
0.785	[['EST' 'LTU' 'LVA'] []]
0.79	[['EST' 'LTU' 'LVA' 'POL' 'RUS'] []]
0.795	[['EST' 'LTU' 'LVA' 'POL' 'RUS'] ['HUN'] []]
0.8	[['EST' 'HUN' 'LTU' 'LVA' 'MEX' 'POL' 'RUS'] []]
0.805	[['EST' 'HUN' 'LTU' 'LVA' 'MEX' 'POL' 'RUS'] ['CZE' 'SWE'] []]
0.81	[['EST' 'HUN' 'LTU' 'LVA' 'MEX' 'POL' 'RUS' 'SWE'] ['CZE'] ['SVK'] []]
0.815	[['CZE' 'EST' 'HUN' 'LTU' 'LVA' 'MEX' 'POL' 'RUS' 'SWE'] ['FIN' 'SVK'] []]
0.82	[['CZE' 'EST' 'HUN' 'LTU' 'LVA' 'MEX' 'POL' 'RUS' 'SWE'] ['FIN' 'SVK' 'TUR'] ['ROU' 'SVN'] []]
0.825	[['CZE' 'EST' 'HUN' 'LTU' 'LVA' 'MEX' 'POL' 'RUS' 'SWE'] ['FIN' 'ROU' 'SVK' 'SVN' 'TUR'] ['ESP'] ['DNK'] ['NOR'] []]
0.83	[['CZE' 'ESP' 'EST' 'HUN' 'LTU' 'LVA' 'MEX' 'POL' 'ROU' 'RUS' 'SWE' 'TUR'] ['DNK' 'FIN' 'ITA' 'NOR' 'SVK' 'SVN'] ['HRV' 'IRL' 'LUX'] ['GBR'] ['BEL' 'DEU'] ['AUT' 'BGR' 'MLT' 'PRT'] []]
0.835	[['CZE' 'ESP' 'EST' 'FIN' 'HUN' 'LTU' 'LVA' 'MEX' 'POL' 'ROU' 'RUS' 'SVN' 'SWE' 'TUR'] ['DNK' 'HRV' 'ITA' 'NOR' 'SVK'] ['BGR' 'DEU' 'GBR' 'IRL' 'LUX'] ['AUT' 'BEL' 'MLT' 'PRT'] []]
0.84	[['CZE' 'ESP' 'EST' 'FIN' 'HUN' 'ITA' 'LTU' 'LVA' 'MEX' 'POL' 'ROU' 'RUS' 'SVN' 'SWE' 'TUR'] ['BGR' 'DEU' 'DNK' 'GBR' 'HRV' 'IRL' 'LUX' 'NOR' 'PRT' 'SVK'] ['AUT' 'BEL' 'MLT' 'TWN' 'USA'] ['CAN' 'FRA' 'NLD'] []]
0.845	[['CZE' 'ESP' 'EST' 'FIN' 'HRV' 'HUN' 'ITA' 'LTU' 'LVA' 'MEX' 'POL' 'ROU' 'RUS' 'SVK' 'SVN' 'SWE' 'TUR'] ['BEL' 'BGR' 'DEU' 'DNK' 'GBR' 'IRL' 'LUX' 'NOR' 'PRT' 'TWN'] ['AUT' 'FRA' 'MLT' 'NLD' 'USA'] ['CAN'] []]
0.85	[['CZE' 'DNK' 'ESP' 'EST' 'FIN' 'GBR' 'HRV' 'HUN' 'IRL' 'ITA' 'LTU' 'LUX' 'LVA' 'MEX' 'NOR' 'POL' 'ROU' 'RUS' 'SVK' 'SVN' 'SWE' 'TUR'] ['AUT' 'BEL' 'BGR' 'DEU' 'FRA' 'MLT' 'PRT' 'TWN' 'USA'] ['CAN' 'NLD'] []]
0.855	[['CZE' 'DNK' 'ESP' 'EST' 'FIN' 'GBR' 'HRV' 'HUN' 'IRL' 'ITA' 'LTU' 'LUX' 'LVA' 'MEX' 'NOR' 'POL' 'ROU' 'RUS' 'SVK' 'SVN' 'SWE' 'TUR' 'TWN' 'USA'] ['AUT' 'BEL' 'BGR' 'CAN' 'DEU' 'FRA' 'MLT' 'PRT'] ['KOR' 'NLD'] []]
0.86	[['BGR' 'CZE' 'DNK' 'ESP' 'EST' 'FIN' 'GBR' 'HRV' 'HUN' 'IRL' 'ITA' 'LTU' 'LUX' 'LVA' 'MEX' 'NOR' 'POL' 'PRT' 'ROU' 'RUS' 'SVK' 'SVN' 'SWE' 'TUR' 'TWN' 'USA'] ['AUT' 'BEL' 'CAN' 'DEU' 'FRA' 'KOR' 'MLT'] ['CYP' 'NLD'] []]
0.865	[['BEL' 'BGR' 'CZE' 'DEU' 'DNK' 'ESP' 'EST' 'FIN' 'GBR' 'HRV' 'HUN' 'IRL' 'ITA' 'LTU' 'LUX' 'LVA' 'MEX' 'NOR' 'POL' 'PRT' 'ROU' 'RUS' 'SVK' 'SVN' 'SWE' 'TUR' 'TWN' 'USA'] ['AUT' 'CAN' 'FRA' 'KOR' 'MLT' 'NLD'] ['CHE' 'CYP'] []]

Table 4 Name of countries failed at that threshold and the cascade impact in 2010

	2010
0.1	[[]]
0.93	[[]]
0.935	[['GRC' 'IRL'] []]
0.94	[['GRC' 'IRL'] []]
0.945	[['GRC' 'IRL'] []]
0.95	[['GRC' 'HRV' 'IRL'] []]
0.955	[['GRC' 'HRV' 'IRL'] ['CYP'] []]
0.96	[['GRC' 'HRV' 'IRL'] ['CYP'] []]
0.965	[['GRC' 'HRV' 'IRL'] ['CYP'] []]
0.97	[['GRC' 'HRV' 'IRL'] ['CYP' 'ESP'] []]
0.975	[['CYP' 'ESP' 'GRC' 'HRV' 'IRL'] ['BGR'] []]
0.98	[['CYP' 'ESP' 'GRC' 'HRV' 'IRL'] ['BGR'] []]
0.985	[['BGR' 'CYP' 'ESP' 'GRC' 'HRV' 'IRL'] []]
0.99	[['BGR' 'CYP' 'ESP' 'GRC' 'HRV' 'IRL'] ['PRT'] []]
0.995	[['BGR' 'CYP' 'ESP' 'GRC' 'HRV' 'IRL'] ['PRT'] []]
1	[['BGR' 'CYP' 'ESP' 'GRC' 'HRV' 'IRL'] ['PRT'] []]

Table 5 Name of countries failed at that threshold and the cascade impact in 2013

	2013
0.1	[[]]
0.85	[[]]
0.855	[['JPN'] []]
0.86	[['JPN'] []]
0.965	[['JPN'] []]
0.97	[['JPN'] ['AUS'] []]
0.975	[['JPN'] ['AUS'] []]
0.98	[['GRC' 'JPN'] ['AUS'] []]
0.985	[['AUS' 'GRC' 'JPN'] ['IDN'] []]
0.99	[['AUS' 'GRC' 'JPN'] ['IDN'] []]
0.995	[['AUS' 'GRC' 'JPN'] ['IDN'] ['TWN'] []]
1	[['AUS' 'GRC' 'IDN' 'JPN'] ['IND' 'TWN'] ['BRA'] []]

Table 6 Name of countries failed at that threshold and the cascade impact in 2015

	2014
0.1	[[]]
0.935	[[]]
0.94	[[]]
0.945	[['JPN'] []]
0.95	[['JPN'] ['AUS'] []]
0.955	[['JPN'] ['AUS'] []]
0.96	[['JPN'] ['AUS'] []]
0.965	[['AUS' 'CYP' 'JPN'] ['RUS'] []]
0.97	[['AUS' 'BRA' 'CYP' 'JPN'] ['RUS'] []]
0.975	[['AUS' 'BRA' 'CYP' 'JPN' 'RUS'] []]
0.98	[['AUS' 'BRA' 'CYP' 'JPN' 'RUS'] ['CAN' 'IDN'] ['TUR'] ['GRC'] []]
0.985	[['AUS' 'BRA' 'CAN' 'CYP' 'JPN' 'RUS'] ['GRC' 'IDN' 'TUR'] ['HRV' 'ITA' 'SWE'] ['AUT' 'EST' 'FIN' 'NLD' 'NOR' 'SVK'] ['BEL' 'CZE' 'DEU' 'DNK' 'FRA' 'LTU' 'LVA' 'MLT' 'SVN'] ['CHE' 'ESP' 'HUN' 'POL' 'PRT' 'ROU'] ['BGR' 'IRL'] ['LUX'] []]
0.99	[['AUS' 'BRA' 'CAN' 'CYP' 'GRC' 'JPN' 'RUS' 'TUR'] ['FIN' 'HRV' 'IDN' 'ITA' 'NOR' 'SWE'] ['AUT' 'CZE' 'DEU' 'DNK' 'EST' 'FRA' 'LTU' 'LVA' 'NLD' 'SVK' 'SVN'] ['BEL' 'CHE' 'ESP' 'HUN' 'MLT' 'POL' 'PRT' 'ROU' 'TWN'] ['BGR' 'IRL' 'LUX' 'ROW'] ['KOR' 'MEX'] []]
0.995	[['AUS' 'BRA' 'CAN' 'CYP' 'GRC' 'HRV' 'IDN' 'ITA' 'JPN' 'RUS' 'TUR'] ['AUT' 'CZE' 'DEU' 'FIN' 'FRA' 'NLD' 'NOR' 'SVK' 'SVN' 'SWE'] ['BEL' 'CHE' 'DNK' 'ESP' 'EST' 'HUN' 'LTU' 'LVA' 'MLT' 'POL' 'PRT' 'ROU' 'TWN'] ['BGR' 'IRL' 'LUX' 'ROW'] ['KOR' 'MEX'] ['GBR'] []]

References

Ahmed, W., Mishra, A. V., & Daly, K. J. (2018). Financial connectedness of BRICS and global sovereign bond markets. *Emerging Markets Review*, *37*, 1–16.

Cabrales, A., Gottardi, P., & Vega-Redondo, F. (2017). Risk sharing and contagion in networks. *The Review of Financial Studies*, *30*(9), 3086–3127.

Degryse, H. A., & Nguyen, G. (2007). Interbank exposures: An empirical examination of contagion risk in the Belgian banking system. *International Journal of Central Banking*, *3*(3), 123–172.

Diebold, F., & Yilmaz, K. (2014). On the network topology of variance decompositions: Measuring the connectedness of financial firms. *Journal of Econometrics*, *182*(1), 119–134.

Elliott, M., Golub, B., & Jackson, M. O. (2014). Financial networks and contagion. *American Economic Review*, *104*(10), 3115–3153.

Helmut Elsinger; Alfred Lehar and Martin Summer. (2006). Risk assessment for banking systems. *Management Science*, *52*(9), 1301–1314.

Mulally, A. R. (2008). Examining the state of the domestic automobile industry, hearing. United States Senate Committee on Banking, Housing, and Urban Affairs.

Stiglitz, Joseph E. (2010). Risk and global economic architecture: Why full financial integration may be undesirable. *American Economic Review*, *100*(2), 388–92.

Trimmer, M. P., Dietzenbacher, E., Los, B., Stehrer, R., & deVries, G. J. (2015). An illustrated user guide to the world input-output database: The case of global automotive production. *Review of International Economics*, *23*, 575–605.

Schweitzer, F., Fagiolo, G., Sornette, D., Vega-Redondo, F., Vespignani, A., & White, D. R. (2009). Economic Networks: The new challenges. *Science*, 422–425.

Esma, N.C. (2006). Solving stochastic PERT networks exactly using hybrid Bayesian networks, In *proceedings of the 7th workshop on uncertainty processing* (pp. 183–197). Oeconomica Publishers.

Kara, F., & Yucel, I. (2015). Climate change effects on extreme flows of water supply area in Istanbul: Utility of regional climate models and downscaling method. *Environmental Monitoring and Assessment*, *187*, 580–596.

Burger, J. D., & Warnock, F. E. (2006). Foreign participation in local currency bond markets. *Review of Financial Economics*, *16*(3), 291–304.

Eichengreen, B., & Luengnaruemitchai, P. (2004). *Why doesn't Asia have bigger bond markets*? NBER Working Paper 10576.

Adelegan, O.J. & Radzewicz-Bak, B. (2009). *What Determines Bond Market Development in Sub-Saharan Africa*? IMF Working Paper No. 09/213, Available at SSRN: https://ssrn.com/abstract=1486531

Mu, Y., Peter, P., & Janet, G. S. (2013). Bond markets in Africa. *Review of Development Finance*, *3*(3), 121–135.

Wasim Ahmad, W., Mishra, A. V., & Daly, K. J. (2018). Financial connectedness of BRICS and global sovereign bond markets. *Emerging Markets Review*, *37*, 1–16.

Timmer, M. P., Dietzenbacher, E., Los, B., Stehrer, R., & de Vries, G. J. (2015). An illustrated user guide to the World Input-Output database: The case of global automotive production. *Review of International Economics*, *23*(3), 575–605.

Dietzenbacher, E., Los, B., Stehrer, R., Timmer, M., & de Vries, G. (2013). The construction of world input–output tables in the wiod project. *Economic Systems Research*, *25*(1), 71–98.